BIOPESTICIDES
Pest Management and Regulation

In memory of Neil Kift

BIOPESTICIDES
Pest Management and Regulation

Alastair Bailey

Kent Business School
University of Kent, Canterbury, UK

David Chandler

Warwick HRI Wellesbourne
University of Warwick, Coventry, UK

Wyn P. Grant

Department of Politics and International Studies
University of Warwick, Coventry, UK

Justin Greaves

Department of Politics and International Studies
University of Warwick, Coventry, UK

Gillian Prince

Warwick HRI Wellesbourne
University of Warwick, Coventry, UK

Mark Tatchell

Department of Biological Sciences
University of Warwick, Coventry, UK

www.cabi.org

CABI is a trading name of CAB International

CABI Head Office
Nosworthy Way
Wallingford
Oxfordshire OX10 8DE
UK

CABI North American Office
875 Massachusetts Avenue
7th Floor
Cambridge, MA 02139
USA

Tel: +44 (0)1491 832111
Fax: +44 (0)1491 833508
E-mail: cabi@cabi.org
Website: www.cabi.org

Tel: +1 617 395 4056
Fax: +1 617 354 6875
E-mail: cabi-nao@cabi.org

© CAB International 2010. All rights reserved. No part of this publication may be reproduced in any form or by any means, electronically, mechanically, by photocopying, recording or otherwise, without the prior permission of the copyright owners.

A catalogue record for this book is available from the British Library, London, UK.

Library of Congress Cataloging-in-Publication Data

Biopesticides : pest management and regulation / Alastair Bailey ... [et al.].
    p. cm.
  Includes bibliographical references and index.
  ISBN 978-1-84593-559-7 (alk. paper)
  1. Natural pesticides--Economic aspects. 2. Natural pesticides–Government policy. 3. Agricultural pests--Biological control. I. Bailey, Alastair. II. Title.

    SB951.145.N37B564 2011
    338.1'62--dc22

2010020101

ISBN-13: 978 1 84593 559 7

Commissioning Editors: Stefanie Gehrig and Rachel Cutts
Production Editor: Tracy Head

Typeset by AMA DataSet, Preston, UK.

Contents

1. Introduction   1

2. Pests of Crops   19

3. Pest Management with Biopesticides   71

4. The Economics of Making the Switch in Technologies   131

5. The Regulation of Biopesticides: an International Analysis   148

6. Policy Networks, Change and Innovation   177

7. Retail Governance   198

8. Conclusions   216

Index   223

1 Introduction

Sustainable Farming and Pest Management: the Great Challenge of the 21st Century

There is no doubt that the need to improve agricultural productivity and enhance its sustainability is one of the most significant challenges facing humanity. Across the world, farmers, growers and others in the food supply chain need to be able to make a fair living from agriculture without damaging the economic, social and environmental prospects of future generations. The scale of the challenge cannot be overstated. Agriculture needs to meet the food requirements of a rapidly expanding global population without placing further pressure on the natural environment and the goods and services it provides. It must adapt to global climate change and mitigate the effects of greenhouse gas emissions, to which it is a major contributor. At some stage, agricultural production – which is heavily reliant on fossil fuels for planting, husbandry, crop protection, harvesting and processing – will have to switch to a low-carbon economy. This will require a radical change in policies.

There is little sign of positive change in agricultural policy; indeed, food production appears to be becoming less sustainable. The global demand for food produced using high-carbon-footprint methods is increasing, driven partly by the rising consumption of meat, while at the same time a succession of poor harvests and expansion of the growth of crops for biofuels are contributing to food price inflation, food insecurity and increasing destruction of natural habitats for farmland. These issues are highly complex, diverse and interconnected. It can be difficult to separate short-term changes in food and farming from long-term trends.

About 40% of the potential global crop yield is destroyed by pests (invertebrates, plant pathogens and weeds) before it is harvested. The diversity of these organisms is set out in Chapter 2. Another 20% is destroyed postharvest. Therefore, making improvements to pest management should be a very significant way of increasing people's access to food. Since the early 1960s, pest management has been heavily reliant on the use of synthetic chemical pesticides. In many respects these have been highly successful tools in raising crop yields. However, as we detail later in this book, the use of conventional chemical pesticides can have significant negative effects for people and the environment, and reliance on them as the sole means of crop protection is not sustainable. Governments around the world are now putting into place policies intended to reduce the use of conventional chemical pesticides.

How can we achieve sustainable agriculture in general and sustainable pest management in particular? The starting point is recognizing that agriculture is multifunctional and provides public goods as they are defined in economics (non-rivalrous and non-excludable): it produces both commodities to be traded on the open market and non-commodities such as natural services, culture and traditions, social fabric and landscape. When done well, agriculture can help provide clean drinking water, provide stewardship for biodiversity and maintain a healthy soil, as well as supplying crops for food, clothing and other materials. The supply of safe and nutritious food and drink is vital for human health, in terms of not only preventing chronic and infectious disease, but also impacting positively on occupational and environmental health. Local foods and employment in their production provide people with a sense of place, purpose and tradition. Pest management plays a central role in all these processes.

However, the standard model of agricultural development has concentrated on increasing commodity production and has produced significant externalities. In the Global North, increases in the production, marketing and sale of highly processed, energy-rich, nutrient-poor foods have contributed to an obesity epidemic. Intensive agriculture has resulted in much loss of biodiversity, degradation of soil and water, air pollution and serious pest management-related problems in the form of resistance, resurgence and invasive species. In overall terms, science and technology have been targeted at improving agricultural productivity at the expense of the environment, social well-being and good health. Things need to change before these problems result in further damage and greater negative feedback into commodity production. We require an effective, sustainable model for agriculture and food that takes full account of the multifunctionality of the supply chain. Unfortunately, as we explore in this book in relation to crop protection and biopesticides, progress is being impeded by gaps in our scientific knowledge, by the way in which agriculture is regulated and by the lack of a market for trading public goods and externalities.

We believe that the natural and social sciences can make a significant positive contribution to the problems facing agriculture in the 21st century. The research on which this book is based formed part of the UK Rural Economy and Land Use (RELU) programme funded by the UK research councils, which facilitated collaborations between social and natural scientists over contemporary challenges in the land-based economies. Our particular concern is with making pest management more sustainable by integrating different control methods according to an ecologically based paradigm. This is known as integrated pest management, or IPM.

Useful definitions of integrated pest management

Prokopy (2003) states that IPM is a 'decision-based process involving coordinated use of multiple tactics for optimizing the control of all classes of pests (insects, pathogens, weeds and vertebrates) in an ecologically and economically sound manner'. In this useful definition, IPM requires a range of techniques to control a range of problems. Within this is an implicit reliance on regular crop monitoring and the careful use of decision rules and thresholds for application of different pest control tactics. Pest management should be done effectively but at the same time minimizing impacts on other components of the agroecosystem, thus taking into account the needs of producers, wider society and the environment (Kogan, 1998). Other workers have placed more emphasis on making use of ecology within the IPM control strategy. Here, Ullyett (1951) makes the claim that IPM is the practice of applied ecology.

IPM does not rule out the use of chemical pesticides. However, in IPM they are not treated as a blanket solution to crop protection but rather are used selectively to reduce their external costs. IPM assumes a broad palette of available control methods. Other IPM-compatible methods include biological, cultural and physical controls, host plant resistance and decision support tools. These are explored further in Chapter 2. The long-term aim is to enhance the functioning of self-regulating ecological systems that occur on farms and which limit the development of pest populations, so that human intervention is applied only when pest populations reach damaging levels. The emphasis in the non-corrective components is that of pest population maintenance.

Biopesticides and other 'alternatives' to chemical pesticides

This book focuses on a particular set of tools for crop protection that have actual or realized potential for farmers and growers. Widely referred to as 'biopesticides' (although this name paints something of a misleading picture, which we discuss later), these are mass-produced, biologically based products for crop protection. They include microbial natural enemies and natural products such as insecticides derived from plants or insect mating pheromones and are described in more detail in Chapter 3. However, to date the sustainability benefits that they offer are not being realized and there are relatively few biopesticides on the market. We discuss some of the ecological and environmental issues concerning biopesticides, the failures in markets and the regulatory process and possible ways forward. We are concerned primarily with the use of biopesticides for the management of plant pathogens, weeds and invertebrate pests of crops in industrialized countries, particularly Europe and North America. Excellent research on biopesticides has been done in these countries, but a lot of pioneering work has also been done in the Global South and we hope we have given this proper recognition in the book.

Agricultural scientists have, for some time, been devoting effort towards the development of alternatives to chemical pesticides in commercial crop production. In a review of alternatives to conventional pest control in the UK, the Advisory Committee on Pesticides (ACP, 2003) identified 14 generic alternative approaches/technologies to conventional chemical pesticides, in addition to monitoring and forecasting techniques, at various stages of development for use in the UK. These approaches are included in Table 1.1.

As Table 1.1 shows, 11 of the generic approaches mentioned by the ACP report are, to some degree, available for commercial use and over 50% of them are assessed to hold reasonable promise as pest management tools. Some of these have achieved high levels of adoption. Good examples include the augmentative introductions of predators and parasitoids for biocontrol of pests in protected crops. In addition, crop rotation, crop breeding and varietal choice are currently used, at varying degrees of intensity, in nearly all annual protected and unprotected crop systems in developed agriculture. However, for the majority, adoption and impact (as measured by reductions in use of conventional chemical pesticides) have been minimal. This lack of commercial adoption could appear to run counter to various assertions by researchers of the appropriateness and even the superiority of these new techniques over the incumbent technology.

Table 1.1. Available range of pesticide alternative approaches. (Adapted from ACP, 2003.)

Technology	Development	Potential
Biological control		
Insect pathogenic fungi	Current	Good
Insect pathogenic viruses	Current	?
Microbial antagonists of plant pathogens	Current	?
Arthropod predators and parasitoids	Current	Good
Management of naturally occurring predators of invertebrate pests	Current	Good
Microbial herbicides	Potential	?
Biochemicals		
Insect pheromones	Current	Good
Antifeedants/eating deterrents	Potential	?
Plant extracts	Near market	Variable
Other methods		
Commodity chemicals	Current	Moderate
Modified atmospheres	Current	Good
Physical and mechanical control	Current	Good
Crop breeding/varietal choice	Current	Good
Rotation	Current	Good

The challenges associated with biopesticides are part of the wider set of issues facing global agriculture that have been set out by the International Assessment of Agricultural Knowledge, Science and Technology for Development (IAASTD). These include:

- improving livelihoods and social welfare for people in the rural economy, particularly for those living in marginalized areas;
- sustaining natural resources, environmental and cultural services, increasing productivity of food, fibre and biofuels, minimizing the negative effects of production and other farm activities; and
- finding effective ways to manage the generation of knowledge and the spread of information.

As we shall explore in this book, the complex interplay between technical issues of pest management, environmental protection, economics, social concerns, regulation and politics that grew out of the chemical pesticide era has had unintended consequences for biopesticides. In the European Union (EU), for example, the regulatory process that resulted first in the large-scale commercialization and adoption of synthetic organic chemical pesticides in the latter half of the 20th century (followed currently by an equally dramatic withdrawal of products from the market) is now acting as a barrier to the commercialization of biologically based products. The great pity is that these biological products can be used in integrated programmes to help reduce the external costs of chemical pesticides while retaining their benefits as efficacious ways of controlling pests. Moreover, the beneficial properties of biologically based agents – such as high specificity (and hence low environmental impact), ease of isolation, potential for self-sustaining control and suitability for cottage production – can be a deterrent to a crop protection industry that has developed along a chemical-based model to focus on high-return, broad-spectrum products and preservation of intellectual property rights.

The focus of this book is on biopesticides and their regulation in comparison to that of chemical pesticides. Some other components of IPM systems are also subject to regulations which may differ from country to country. In particular, the introduction and release of exotic natural enemies for biological control are frequently subject to environmental legislation designed to prevent the introduction of damaging organisms into new environments. This subject is not discussed further here.

The Regulatory State

A central theme of this book is the existence of regulatory barriers to the wider use of biopesticides which could contribute to IPM strategies. An alternative paradigm to regulatory failure is that of market failure, which, put at its simplest, means that some products are insufficiently technologically advanced or reliable in their effects or face a lack of market demand to make them commercially viable given the cost of their research, development and registration. There is no doubt that this is a real challenge for the biopesticides industry, as although there are a few broad-spectrum products, most of those developed and available on the market tend to be niche products with specific applications to particular crops that are not produced on an extensive scale. They are often referred to as 'minor crops' in contrast to broad-acre crops grown on a large scale, although this term is somewhat misleading as a contribution to the policy debate as it covers significant fruit and vegetable crops such as apples and brassicas.

We do not want to underplay the significance of these economic considerations and Chapter 4 is written from an economics perspective. There is no doubt that the industry has from time to time failed to deliver what it promised and this has been damaging to its reputation. A system of registration is needed to discourage 'snake oil' products that cause reputational damage and, indeed, there is a case for more rigorous regulation of so-called 'grey market' products, which do not make explicit pest control claims and are therefore generally outside the scope of the regulatory system, but may be used as a means of pest control.

A central initial hypothesis in our project was the notion of 'regulatory failure'. It should be emphasized that this did not mean that regulatory agencies or individual regulators had been incompetent or had failed to apply regulatory rules properly. Instead it refers to the fact that the regulatory system had been designed for synthetic products and had difficulty in responding to the different characteristics and properties of biological products. Product developers, typically small firms, were discouraged from applying for registration at all or found the process unduly lengthy, onerous and costly. As will be discussed in the book, regulatory agencies, especially the Pesticides Safety Directorate (PSD) (now the Chemicals Regulation Directorate, CRD) in Britain, have attempted to respond systematically to these challenges and to make the regulatory system more responsive to biocontrol products.

Our general approach is framed within a widely used paradigm in political science known as the 'regulatory state', but our more specific analysis uses 'regulatory innovation', discussed below. The regulatory state is an ideal typical model, i.e. it does not apply to all aspects of political reality, but is used to capture a movement from the early Keynesian welfare state (substantial elements of which remain) to a new emphasis on regulation as a mode of governance. Issues raised by the regulatory state debate are relevant to our consideration of pesticides regulation. 'Pesticides are amongst the most strictly regulated of all chemicals' (House of Commons, 2005: 6). Models of the regulatory state, therefore, provide a framework to our discussion. Moran's work (2000, 2002, 2003, 2005) is of particular importance in terms of developing a model of its essential characteristics. New executive agencies are contracted to deliver policy; a newly privatized sector is subject to a network of specialized regulatory agencies; and government has turned to the specialized regulatory agencies to control large areas of economic and social life (Moran, 2005: 156).

The regulatory state model has been primarily used in relation to the developed world and Phillips (2006) has justifiably criticized its lack of applicability to the Global South. However, for our particular purposes, namely a geographical focus on Europe and the USA (together with Australia and Canada), the model has explanatory value. Its utility may be limited to the developed world but it can still have utility in comparing modes of governance within a particular state. On that particular point our perspective diverges from the argument made by Phillips that the regulatory state research agenda offers 'little purchase on the elements of different varieties of capitalist organisation, policy orientation and institutional design that exist within national political economies' (Phillips, 2006: 22).

In his initial work on the regulatory state, Moran took a relatively benevolent view of the regulatory stage, which in popular discourse might be seen to refer to what is sometimes called 'the nanny state'. He saw it as a great improvement on earlier state forms. As he puts it:

> As for the old world of command, good riddance to it: good riddance to the men in Whitehall who know best ... The world of command infantilised us all – never let us grow up from subjects to citizens ...
>
> (Moran, 2000: 12)

Indeed, it is possible to read Moran's work as an interpretation of a progressive, yet imperfect and complete, transition towards modernity. In ideal typical terms, the 'command' or 'Keynesian welfare state' is displaced, partly because of the exhaustion of this earlier paradigm, by the regulatory state. This involves more indirect forms of state control which may, of course, reinvigorate rather than reduce state power (Wolfe, 1999). Progressive features of the new dispensation include the displacement of 'club' government and the replacement of self-regulation which was seen to fail both in terms of economic efficiency and public accountability. In the case of pesticides, for example, voluntary regulation under the Pesticides Safety Precaution Scheme in the UK was displaced by statutory regulation in 1986.

Moran has become more impressed by the regulatory state's authoritarian potential (M. Moran, personal communication). He suggests we look to the authoritarian strand in regulation; his recent textbook argues that *The British Regulatory State* (Moran, 2003) recognized this strand and 'paints the British regulatory state in a threatening and interventionist light' (Moran, 2005: 530). Moran now accepts that this remark does not quite gloss it accurately (Moran, 2005: 530). Nevertheless, he explains how his 2003 work argued that the regulatory state had a Janus face: a democratizing quality, because it enforces more transparency on elites, but also an authoritarian quality, because it also centralizes and controls. To complicate matters further, the latter feature encourages it in the direction of failures and catastrophes, which, in turn, subverts its control capabilities (Moran, 2005: 530). He argues that 'the British regulatory state, far from being smart, is, therefore, often remarkably stupid'. He adds, however, that 'it succeeded a governing system that was even more stupid' (Moran, 2003: 26).

One recurrent theme in the literature on regulation is the danger of 'regulatory capture'. The new regulatory state sets up relationships between regulator and regulated that are close, involving daily contact. There is an obvious danger, therefore:

> that the regulator will grow so like the regulated that the two will simply share common interests and a common view of the world, and the independence so vital to the new regulatory agencies will be undermined.
>
> (Moran, 2005: 159)

This is a strong theoretical tradition in the USA that reflects a system of government in which historically there have been 'iron triangle' relationships between congressional committees, regulatory agencies and the regulated who are often substantial donors to the re-election campaigns of the members of the congressional committees. There certainly can be problems of asymmetric information between regulators and the regulated and there is some evidence to suggest this has been a problem in utility regulators in the UK (Willman et al., 1999; Héritier, 2005). We have found little evidence of it within PSD. The organization has built up its own in-house scientific expertise so that it can make independent and authoritative judgements on pesticide approvals. It seems to be an agency that is relatively insulated from external opinion and, as we discuss later, a combination of exogenous intervention and internal responsiveness to it was necessary to create a regime more amenable to biopesticide registration.

Regulatory Innovation

The 'reinventing government' movement in the 1990s was a response, on the one hand, to the excessive faith in 'big' government that had predominated from the New Deal of the 1930s to the 1970s and the over-reaction to it in terms of the belief propagated by Ronald Reagan and Margaret Thatcher in the 1980s that government was part of the problem not part of the solution. It was felt that what was needed was a form of government that was both leaner and smarter and more targeted and focused in its use of limited resources. The notion of innovation is a key part of the 'reinventing government' debate (Osborne and Gaebler, 1993) and has a key role in debates on regulatory reform (e.g. OECD, 1995; European Commission, 2002). This links to notions of governance that emphasized that government would 'steer' rather than 'row' (Rhodes, 2000) and also the desire for 'smarter', more effective regulation. Public agencies were told that innovation should become one of their 'core activities' (Cabinet Office, 2003). Innovation studies in political science date back to the 1960s but there has been a revival of interest in studying specific examples of policy innovation in the delivery of public services by governmental and voluntary organizations (e.g. Doig, 1997; Light, 1998; Borins, 2000). There is now also a web-based journal devoted to the topic (http://www.innovation.cc). The Better Regulation Executive (BRE) operates within the Department for Business, Innovation and Skills (BIS) and leads the regulatory reform agenda across government. Such structures could be expected to be re-branded by any change of government.

Regulators are typically risk averse. The consequences of making a mistake are serious, not least where public safety and environmental protection are involved. This does not create an encouraging environment for regulatory innovation: indeed, the term is almost a contradiction. This book will argue that regulatory innovation has also successfully occurred within CRD (see Chapter 6). At a basic level regulatory innovation is often needed to offset the problems of regulatory failure which we referred to earlier. More specifically, in terms of the regulatory state model, the priority given to regulation sets up expectations of innovation and responsiveness to societal demands for change which can be hard to meet in practice. There is a fundamental tension, moreover, between expectations that regulators will be consistent, predictable and impartial, and yet also innovative.

One image is that regulatory innovation is a matter of refining the technologies of regulation: again, the search for better tools of governance, the development of 'smart regulation' (e.g. Osborne and Gaebler, 1993; Sparrow, 2000). For others, regulatory innovation is seen as necessary in order to improve strategies for managing risk, as seen, for instance, in the rise of 'new public risk management' (e.g. Cabinet Office, 2000; Sparrow, 2000). In this image, regulatory innovation is the pursuit of state legitimacy in the risk society (Beck, 1992). At a more theoretical level, regulatory innovation is an adaptive response by a system or organization to its environment. The problem with such images is that they have conflicting normative and cognitive implications (Black, 2005a: 4). Black, therefore, offers a normative neutral definition which we also propose to use:

> Regulatory Innovation is understood ... to be the use of new solutions to address old problems, or new solutions to address 'new' (or newly constructed) problems, but not old solutions to address old problems.
>
> (Black, 2005a: 4)

This is illuminated by Hall's (1993) typology of policy change. First-order changes are changes to the levels or settings of basic policy instruments (not considered here as innovations). Second-order policy changes involve changes in technique, process or instrument, but not in the overall goals of policy or understanding on which it is based. Third-order changes involve changes in the goals of policy and understandings on which the policy is based, along with second- and first-order changes. Third-order changes are 'paradigm shifts', changes in the terms of a policy discourse: the understandings on which it is based and the goals pursued (Kuhn, 1962).

The dominant assumption in the literature is that innovation is 'a good thing'. In Mohr's influential definition, innovation is seen as the 'successful introduction into an applied situation of means or ends that are new to that situation' (Mohr, 1969: 112). Innovation was defined by the Cabinet Office, moreover, as 'new ideas that work' and more precisely as 'the creation and implementation of new processes, products, services and methods of delivery which result in significant improvements in outcomes, efficiency, effectiveness or quality' (Cabinet Office, 2003: para 2.1). In contrast, Moran draws attention to some of the errors that may result from too great an emphasis on innovation (Moran, 2003). Clearly, innovation need not always be successful. Indeed, being in a constant state of innovation can be counterproductive: initiatives may not be given the time to be properly implemented; costs may be imposed through the need to change systems and processes; and 'no policy is around for long enough for its success or failure to be properly accessed' (Black, 2005a: 14).

So how and why does regulatory innovation occur? Black (2005b) argues that the different sets of explanations tend to occupy different 'worlds' or ways of thinking. The outline of each world is not intended to be a complete account of the theories of innovation that may potentially exist or be grouped within those worlds, in particular because the literature which specifically refers to innovation is itself quite narrow and does not necessarily seek to engage with wider questions relating to policy change. The worlds identified are fivefold: the world of the individual, the organization, the state, the global polity and the innovation. In the world of the individual the individual is both the site and agent of innovation. The organizational world concentrates on innovations within organizations, traditionally profit-making organizations but increasingly non-profit public and private sector organizations. The state world's site of analysis is the public policy-making processes of the state. The global polity looks at policy making by global polity bodies and networks. Finally, the world of the innovation revolves around the idea of innovation itself. We look at these 'worlds' in more detail in Chapter 6. In our particular case, however, a consideration of the contextual and exogenous and endogenous drivers (while overlapping with Black's analysis) will be shown to provide a better framework. Furthermore, we will outline a more detailed model to account for regulatory innovation in regulatory agencies.

The Interdisciplinary Challenge

The policy challenges faced in today's world often require disciplines that had not collaborated very much in the past to work together to undertake analysis and develop policy solutions. In particular, there is an imperative for economics and political science to collaborate not just with other social sciences, where there is a significant amount of relevant experience, but also with the natural sciences, a territory that is less well mapped and explored. Such an imperative is not just policy related, but also academically driven.

In our work on biopesticides we have sought to move beyond multidisciplinarity in the sense of research that 'requires scholars to be aware of salient contributions from other disciplines and to draw on them in useful and limited ways' (Warleigh-Lack and Cini, 2009: 8). We have sought to move towards interdisciplinarity involving 'a much more sustained process of dialogue, together with joint problem-definition, and methodology, most probably devised by a team of scholars from the salient disciplines and areas' (Warleigh-Lack and Cini, 2009: 8–9). Probably the greatest challenge is to develop a common methodology, and we would not claim to have achieved that. We have not aspired to or achieved transdisciplinarity by 'adding an overarching common meta-theoretical perspective to the common definition of the problem and methods' (Warleigh-Lack and Cini, 2009: 9).

It is possible that working with cognate disciplines in the social sciences may in some cases be more difficult than working with natural scientists, in part because there may be contested boundaries and fears about capture. The closer the two disciplines, the greater this fear may be. Because the gap is wider with natural science, it may be more challenging to bridge, but there may be fewer fears about leaping the chasm and the rewards of doing so may be greater.

The natural sciences and the social sciences

There are some evident differences between the natural sciences and the social sciences. The comparison here is limited to biological science and political science. Different issues may arise in relation to, say, physics, where it is at least arguable that a great deal of meta-theoretical speculation takes place that is not experimentally verifiable (Smolin, 2005, but his views are contested).

In this discussion the focus is on four differences which may not always be as substantial as they initially appear to be:

1. The Oedipus effect.
2. The greater ease of experimentation within the natural sciences.
3. The use of highly sophisticated systems to understand problems.
4. The possibility of generalizing from an individual organism or species without committing an individualistic fallacy.

The Oedipus effect
Science is about explanation, but one key test of its worth is its ability to make sustainable and verifiable predictions. If one predicts that a material will expand at a given rate when subjected to heat, that hypothesis can be tested and verified and it can influence the way in which the material is used in a variety of situations. In social science, a prediction may affect or even change the outcome. For example, supposing a credible source predicts that Party A is going to win a substantial majority over Party B in an election and that the gap cannot be closed. Supporters of Party B could be discouraged from voting because they think there is little point, or supporters of Party A could decide that the election is already won and they do not need to vote. In either case (or a combination of both of them) the outcome is affected, albeit in somewhat complex or unpredictable ways.

Experimentation
An important difference between biological science and political science would appear to be the ease with which controlled experiments can be

undertaken in biology. 'Classically, what makes an experiment is [the] distinction between the control situation and the *experimental* situation.' In the experimental situation, 'the variable of interest is changed to some value at which we hope to demonstrate causality' (Cohen and Medley, 2005: 44).

Much of experimental biology in recent decades has moved away from the interpretation of associations towards the understanding of causal relationships in dynamic systems. This has been enhanced by the huge advances in molecular biology and genetics. Biological science often relies on simplified systems for study. These often consist of small numbers of experimental units maintained under controlled conditions. This does raise an issue whether something that can be observed in the highly controlled conditions of the laboratory can be replicated elsewhere. Forms of complexity may be largely hidden in idealized laboratory settings. Csete and Doyle (2002) explain the concept of spiralling complexity in terms of the need for complex regulatory networks to ensure robustness. They use the example of a Lego model made mobile, then motorized and finally equipped to avoid collision in a maze of obstacles, this third step increasing complexity by orders of magnitude. 'This is consistent with the claim that biological complexity too is dominated not by minimal function, but by the protocols and regulatory feedback loops that provide robustness and evolvability' (Csete and Doyle, 2002: 1666). Much of this complexity is not immediately apparent to the observer. Nevertheless, in order to understand complex phenomena, it is often necessary to take a reductionist/simplified approach. One does not want to replicate complexity in any scientific undertaking, but to offer a parsimonious model of it.

One should not push the notion that science is dominated by manipulative experiments too far. While a manipulative experiment may be the ideal, the practical difficulties mean that they are often impossible, or restricted to small sample sizes or unrealistic simplified circumstances. Plentiful observational data gathered in natural circumstances 'in the field' are often preferable. In ecology and epidemiology much work is based on the statistical analysis of observational data.

The use of highly sophisticated systems
The study of elections has becoming increasingly sophisticated in its use of quantitative techniques, taking advantage of advances in econometrics. Nevertheless, this cannot be compared with the advances made in systems biology. This approach in cell and molecular studies uses mathematics and sophisticated computing to understand highly complex systems, based on integrating large data sets collected from many experiments done at the individual level. What it makes possible is an understanding of the structure and dynamics of cellular and organism function, rather than the characteristics of the isolated parts of a cell or organism. In particular, progress 'in genome sequencing and high-throughput measurements, enables us to collect comprehensive data sets on system performance and gain information on the underlying molecules' (Kitano, 2002: 1662). It is still a work in progress and requires improvements in software infrastructure and 'high throughput and

accurate measurements, goals that are perhaps beyond the scope of current experimental practices' (Kitano, 2002: 1663).

Systems biology places considerable emphasis on both system structures and system dynamics. However, 'In reality, analysis of dynamics and structure on the basis of network dynamics are overlapping processes' (Kitano, 2002: 1662). In political science an analogy can be found in the substantial body of work on policy networks and communities which we use extensively in this book. This analysis attempts to define the boundaries of such networks, how they interact (including subsystems where relevant) and how this leads to particular outputs such as legislation or institutional design. However, in political science, this is very much a meso level rather than a systems level analysis. It does not have the transformative impact of systems biology where there is a transition 'from the molecular level to the system level that promises to revolutionize our understanding of complex biological regulatory systems' (Kitano, 2002: 1662). Can political science be as ambitious as biological science? Does it have the tools to match ambition with appropriate and rigorous analytical techniques?

'Robustness is an essential property of biological systems' (Kitano, 2002: 1663). This claim has a relevance to political systems, particularly in terms of two of the three properties exhibited by robust systems. The first of these is 'adaptation, which denotes the ability to cope with environmental changes' (Kitano, 2002: 1663). Political systems have to be able to adapt to changes in their wider environment if they are to survive and prosper. For example, domestic political systems need to be able to cope with the challenges of globalization, such as by grouping together in new ecologies at a regional level.

An important property of a political system could also be 'graceful degradation, which reflects the characteristic slow degradation of a system's function after damage, rather than catastrophic failure' (Kitano, 2002: 1663). Subjected to an external shock, a robust political system should not collapse completely. A slow degradation allows new structures and functions to replace it successfully.

Parameter insensitivity as a property of robust systems poses more challenges for analysis. This is because the boundaries of political systems are relatively contested and often very permeable. The boundaries are shifting, not just geographically, but in terms of social and economic formations; indeed, this permeability increases under conditions of globalization. This is the case both geographically and in terms of who is allowed to constitute a citizen of the polity, an issue that becomes particularly potent at the interface of migration and globalization. This observation raises the broader issue that the very terms we use in political science are essentially contested because they have a substantial normative component.

The individualistic fallacy
Both the biosciences and political science face the risk of committing the ecological fallacy and the individualistic fallacy. The ecological fallacy entails 'inferring without investigation that relationships among collectivities are

the same as those for individuals. Only under very specific circumstances ... are such inferences from ecological data valid; otherwise, the observer has committed the ecological fallacy' (Alker, 1965: 102). In other words, the ecological fallacy involves the identification of statistical relationships at the aggregate level that do not accurately reflect the corresponding relationship at the individual data level. 'Anyone who *draws a conclusion about individuals based on evidence about groups* commits what is called the "ecological fallacy"' (McIntyre, 2005: 42).

'It is likewise a logical error to draw conclusions about groups based on data gathered with the individual as the unit of analysis' (McIntyre, 2005: 43). This is known as the individualistic fallacy or sometimes as the reductionist fallacy. 'The individualistic fallacy is just the opposite of the ecological fallacy ... social scientists are ... likely to try to generalize from individual behaviour to aggregative relationships' (Alker, 1965: 103). In biology, an individualistic fallacy can be committed in two ways: (i) by making inference about a group from data collected on an individual of the same species; and (ii) by making inference about one species from data collected on another species.

How is it possible for bioscientists to generalize on the basis of observations on an individual organism or species without committing an individualistic fallacy? One approach that is used is the use of model systems/organisms. These are chosen because they are easy to work with, but they still have relevance to less tractable world systems. Examples of model organisms include the lab rat and fruit fly. In freshwater biology, the 'crustacean *Daphnia magna* Straus is an excellent model organism for stage-structured demographic analysis. Under favourable conditions, females reproduce parthenogenetically, producing large broods of female offspring at intervals of 3–4 days' (Carslake *et al.*, 2009: 1077). There is also the model plant, a widely used example being *Arabidopsis thaliana*, a non-commercial member of the mustard family. Instead of studying many different plants, one can study this particular plant that has a number of helpful characteristics for research purposes (although the similarity of *Arabidopsis* to other flowering plants may be a little simplistic and there have been real difficulties in transferring the findings from *Arabidopsis* even to the closely related brassica crops). Nevertheless, it has real advantages:

- it is easy and inexpensive to grow;
- it produces many seeds;
- it responds to stress and disease in the same way as most crop plants;
- it has a small genome (genetic complement) facilitating genetic analysis.

Use of the model plant is possible because all flowering plants are closely related. Complete sequencing of the genes of a single, representative plant will yield knowledge about all higher plants. Biological material that is genetically identical can be generated through carefully designed crossing and backcrossing programmes that produce isogenic lines. This material is then used to study causal relationships.

Because only one organism/system is studied by many people, resources are pooled and knowledge is acquired rapidly. The degree to which a scientific observation made on the model organism applies to a different species (i.e. the probability of committing an individualistic fallacy) depends on the relationship between the two species and the scientific hypothesis that is being tested. One of the attractive properties of model organisms/systems is that they provide a baseline or anchor point. You can test a hypothesis in the tractable, well-studied model organism, and then see if it applies to other organisms. Human behaviour is much more diverse. We cannot identify a 'model citizen' from whom we generalize. (Interestingly, the media does sometimes seek to identify the median voter: in the UK, 'Worcester woman', a woman in her 30s living in the city of Worcester, with children, in part-time work and driving a Ford Mondeo.)

What is not being argued is that biological science is superior to political science in terms of its predictive and explanatory power simply because of its subject matter and the research techniques available to it. Although biological systems are in one sense less complex than politics and so are easier to explain and predict, at a fundamental level the question of 'superiority' is an unhelpful distraction, particularly when one is trying to foster collaboration. What is needed in both biological and political science, and especially in joint projects between them, is to develop the best methodologies available in order to understand the system under study in the most complete yet also parsimonious way. In any field of study, the most complex and possibly intractable systems are those that cannot be described mathematically, but which have to be described and analysed using words.

Complementarities and differences between the natural and social sciences
The Oedipus effect captures a real difference between the natural and social sciences in terms of the fact that the social sciences are dealing with human agents with a capacity to learn and reflect, who may change their behaviour in response to predictions. The use of experimental methods is growing in political science, but it is still at the periphery of the discipline compared with its central place in many of the natural sciences. For many of the advocates of interdisciplinarity, 'the appropriate focus of interdisciplinary study is on specific complex systems and their behaviour' (Newell, 2001: 2). For political scientists who favour a systems level approach, interdisciplinarity could offer a welcome refuge. The discussion of the individualistic fallacy found that biological scientists did not necessarily have more reliable ways of overcoming it than political scientists.

Conclusions

The food chain faces the challenge of producing enough food sustainably against the background of a growing world population and increased per capita demand for food, and for food that demands more resources to produce, as emerging countries such as China and India become more prosperous. We face simultaneously a food security challenge and a sustainable production challenge. An important part of that sustainability challenge is ensuring a response to climate change mitigation and adaptation, a theme to which we return in the conclusions (Chapter 8). In emphasizing climate change we should not forget that other environmental challenges remain such as pollution control and maintaining biodiversity.

IPM offers a strategy for providing crop protection while tackling environmental challenges and consumer concerns. Biocontrol products, of which biopesticides are a subset, form an important component of an IPM approach. They have primarily been niche products mainly deployed on high-value crops grown on small areas rather than broad-acre crops. In the case of broad-acre crops, a range of agrochemical solutions is available. For many so-called minor crops, including widely eaten field vegetables, the number of crop protection solutions is diminishing rapidly and the availability of suitable biocontrol products is an urgent need. The development of new products continues and sales are growing significantly, with big companies starting to buy back into the sector for the first time since the 1980s. Among the drivers here are the new EU legislation which we discuss in Chapter 5, the availability of broader-spectrum products, the illegal synthetics scandal in Spain and the concerns of consumers about synthetics refracted through retailers (see Chapter 7).

The sector has faced market failure problems in terms of whether the size of the market for some products is sufficiently large to recoup research and development costs. Some products have not lived up to the marketing claims made about them. However, there has also been a regulatory failure problem and this is a particular focus in this book. This requires the design of appropriate policy instruments which may range from broad-gauge agri-environmental schemes (discussed in Chapter 4) to measures designed to promote the regulation of biocontrol agents (see Chapter 5). It also requires thinking about the scope of regulation in terms of 'grey market' products which may undercut products that have incurred the costs of registration. Many challenges remain for IPM and biocontrol, but there is also an unparalleled window of opportunity within the EU.

References

ACP (2003) *Final Report of the Sub-group of the Advisory Committee on Pesticides on Alternatives to Conventional Pest Control Techniques in the UK: a Scoping Study of the Potential for Their Wider Use*. Advisory Committee on Pesticides, York, UK.
Alker, H.J. (1965) *Mathematics and Politics*. The Macmillan Company, New York.
Beck, U. (1992) *Risk Society: Towards a New Modernity*. Sage, London.
Black, J. (2005a) What is regulatory innovation? In: Black, J., Lodge, M. and Thatcher, M. (eds) *Regulatory Innovation*. Edward Elgar, Cheltenham, UK, pp. 1–15.
Black, J. (2005b) Tomorrow's world: frameworks for understanding regulatory innovation. In: Black, J., Lodge, M. and Thatcher, M. (eds) *Regulatory Innovation*. Edward Elgar, Cheltenham, UK, pp. 16–44.
Borins, S. (2000) Loose cannons and rule breakers, or enterprise leaders? Some evidence about innovative public managers. *Public Administration Review* 60, 498–507.
Cabinet Office (2000) *Successful IT: Modernising Government in Action*. HMSO, London.

Cabinet Office (2003) *Innovation in the Public Sector*. HMSO, London.

Carslake, D., Townley, S. and Hodgson, D.J. (2009) Predicting the impact of stage-specific harvesting on population dynamics. *Journal of Animal Ecology* 78, 1076–1085.

Cohen, J. and Medley, G. (2005) *Stop Working and Start Thinking: a Guide to Becoming a Scientist*, 2nd edn. Taylor & Francis, Abingdon, UK.

Csete, M.E. and Doyle, J.C. (2002) Reverse engineering of biological complexity. *Science* 295, 1664–1669.

Doig, J.W. (1997) Leadership and innovation in the administrative state. *International Journal of Public Administration* 20, 861–879.

European Commission (2002) *Innovation Policy and the Regulatory Framework*. European Commission, Brussels.

Hall, P.A. (1993) Policy paradigms, social learning, and the state: the case of economic policymaking in Britain. *Comparative Politics* 25, 275–296.

Héritier, A. (2005) Managing regulatory developments in rail compliance and access regulation in Germany and the UK. In: Coen, D. and Héritier, A. (eds) *Refining Regulatory Regimes: Utilities in Europe*. Edward Elgar, Cheltenham, UK, pp. 120–144.

House of Commons (2005) *Progress on the Use of Pesticides: the Voluntary Initiative*. Eighth Report of Session 2004–5. Environment, Food and Rural Affairs Committee. HMSO, London.

Kitano, H. (2002) Systems biology: a brief overview. *Science* 295, 1662–1664.

Kogan, M. (1998) Integrated pest management: historical perspectives and contemporary developments. *Annual Review of Entomology* 43, 243–270.

Kuhn, T. (1962) *The Structure of Scientific Revolutions*. Chicago University Press, Chicago, Illinois.

Light, P. (1998) *Sustaining Innovation*. Jossey-Bass, San Francisco, California.

McIntyre, L.J. (2005) *Need to Know: Social Science Research Methods*. McGraw Hill, New York.

Mohr, L. (1969) Determinants of innovation in organizations. *American Political Science Review* 75, 111–126.

Moran, M. (2000) From command state to regulatory state. *Public Policy and Administration* 15(4), 1–13.

Moran, M. (2002) Review article: understanding the regulatory state. *British Journal of Political Science* 32, 391–413.

Moran, M. (2003) *The British Regulatory State: High Modernism and Hyper-Innovation*. Oxford University Press, Oxford.

Moran, M. (2005) *Politics and Governance in the UK*. Palgrave Macmillan, Basingstoke, UK.

Newell, W.H. (2001) A theory of interdisciplinary studies. *Issues in Integrative Studies* 19, 1–25.

OECD (1995) *Regulatory Reform and Innovation*. Organisation for Economic Co-operation and Development, Paris.

Osborne, D. and Gaebler, T. (1993) *Reinventing Government*. Plume, New York.

Phillips, N. (2006) States and modes of regulation in the global political economy. In: Minogue, M. and Carino, L. (eds) *Regulatory Governance in Developing Countries*. Edward Elgar, Cheltenham, UK, pp. 17–38.

Prokopy, R.J. (2003) Two decades of bottom-up, ecologically based pest management in a small commercial apple orchard in Massachusetts. *Agriculture, Ecosystems & Environment* 94, 299–309.

Rhodes, R.A.W. (2000) The governance narrative: key findings and lessons from the ESRC's Whitehall program. *Public Administration* 78, 345–363.

Smolin, L. (2005) *The Trouble with Physics*. Penguin, London.

Sparrow, M.K. (2000) *The Regulatory Craft: Controlling Risks, Solving Problems and Managing Compliance*. Brookings Press, Washington, DC.

Ullyett, E.C. (1951) Insects, man and the environment. *Journal of Economic Entomology* 44, 459–464.

Warleigh-Lack, A. and Cini, M. (2009) Interdisciplinarity and the study of politics. *European Political Science* 8, 4–15.

Willman, P., Coen, D., Currie, D. and Siner, M. (1999) *Regulatory Organisation and Regulatory Effectiveness in Privatised Companies. Regulation Initiative Discussion Paper Series No. 22*. London Business School, London.

Wolfe, J.D. (1999) Power and regulation in Britain. *Political Studies* 47, 890–905.

2 Pests of Crops

A pest is any organism that reduces the availability, quality or value of a human resource including our crop plants (Flint and van den Bosch, 1981). Crop pest organisms include plant pathogens (fungi, oomycetes, bacteria, viruses, nematodes and some other taxa), weeds, invertebrates (primarily insects, mites and molluscs) and a small number of vertebrate species. Each species and variety of crop plant is exploited by a community of pest species, the composition of which varies according to country and region. Rice, for example, is susceptible to more than 20 major insect pests and plant pathogens (Bonman *et al.*, 1992), while oilseed brassica crops are attacked by about 50 major insect pests and plant pathogens (Lamb, 1989; Leino, 2007). Overall, there are estimated to be about 67,000 different pest species and they are a significant constraint on agricultural production, responsible for around 40% loss of potential global crop yields (Oerke *et al.*, 1994; Pimentel, 1997). Of this, 15% is caused by arthropods, 12–13% by plant pathogens and 12–13% by weeds. A further 20% loss in yield is estimated to occur after crops have been harvested and placed into storage. These losses occur despite the very considerable efforts made at pest control and they suggest that improvements in pest management are likely to be a significant route forward for improving yields and access to food in the years ahead (Oerke and Dehne, 2004). About 90% of the world's food energy intake comes from only 15 species of crop plant, and 66% from just three species: rice, maize and wheat (FAO, 2010). If our crop protection systems for any one of these crops were to fail, then the consequences could be severe. The loss potential of pests worldwide has been estimated to vary from less than 50% on barley to more than 80% on sugarbeet and cotton (Oerke and Dehne, 2004). Famines and starvation due solely to pests are mercifully rare, but they have occurred in the past. Probably the best-known example, the Irish potato famine of 1846, caused mass starvation and the creation of the Irish Diaspora. More commonly, the effects of crop pests combine with other constraints on crop production (weather, soil fertility, labour, farmer access to knowledge, market prices and government policies on agriculture) to cause food insecurity (Speranza et al., 2008). As the pressure on our food supply increases over the next decades, then crop pests are bound to become more of a concern for global food supplies.

In nature, consumption of plants is a fundamental component of virtually every terrestrial food web, and hence we should not be surprised that there are plenty of organisms 'out there' that are able to feed on our crops. But what is often not fully appreciated is that the creation and management of agricultural land cause the conditions under which pests flourish. Growing crops in monoculture provides a concentrated food resource that allows phytophagous pest populations to achieve far higher densities than in natural environments, while clearing the ground by ploughing creates ideal conditions for the growth of weeds. All plant life stages and all portions of the plant are vulnerable to pests. Reduction in yield or shortfall in crop quality occurs because of consumption of plant nutrients by pathogenic microorganisms and herbivorous animals or by competition for resources with agricultural weeds. The different mechanisms by which pests affect the crop have evolved as a result of natural selection and are driven by competitive interactions between the pest and the plant, between different pest species (e.g. for food and space), with other members of the ecological community (e.g. with predators or disease) and the abiotic environment. Crops are attacked by both generalists, which affect many plant species, and specialists, which feed on a narrow spectrum of plant species and show specific adaptations to plant defences or certain habitats. Pest damage can be direct (i.e. the plant is eaten by a pest) or it can be indirect, in which case there is a reduction in yield or quality due to competition for resources (most weeds reduce crop yields in this way) or because the pest acts as a vector of a plant disease. In general, seedlings and young plants are more vulnerable than mature plants. Direct damage is most serious when it occurs to the harvestable part of the plant. Plants can also exhibit compensatory growth in response to pest attack and hence a crop may be able to tolerate partial loss of leaves without a reduction in yield.

Biology of the Main Groups of Agricultural Pests

Describing the biology, ecology, behaviour and function of the main agricultural pests in detail is beyond the scope of this book, and indeed many books have been devoted to this task. For more information we recommend Hill (1994), Naylor (2002) and Agrios (2005). Here we provide a brief introduction to pest biology as a foundation for later sections in the book on biopesticides and sustainable agriculture.

Invertebrates

Insects
Insects are among the most abundant and diverse groups of animals on the planet. There are estimated to be 6–10 million extant species and they make

up about half of all species of higher organisms (Gullan and Cranston, 1994). There are about 30 insect orders, of which ten orders and over 50 families are considered to contain major pests of crops (Hill, 1994, 1997). These include the Orthoptera (pests in this order include locusts, grasshoppers and crickets), Hemiptera (sap-feeding insects that include a wide variety of pests such as aphids, whitefly, plant hoppers, leaf hoppers, scales and capsid bugs), Thysanoptera (thrips), Coleoptera (beetles and weevils), Lepidoptera (butterflies and moths), Diptera (root and stem flies) and Hymenoptera (sawflies). A list of some of the most important members of these taxonomic groups is given in Table 2.1. Most insect species have evolved specialized mouthparts for biting, sucking or chewing, and immature and adult stages of agricultural pests use these to attack different parts of the host plant. The hard, waxy exoskeleton of insects means that many species are resistant to desiccation and can survive in harsh conditions. Some of the most important insect pests cause indirect damage by vectoring plant pathogens. Aphids (Aphididae), leaf hoppers (Cicadelidae) and plant hoppers (Delphacidae) are all vectors of plant viruses, while some beetle (Coleoptera) pests are vectors of fungal

Table 2.1. Insect pests of agricultural crops. (From Hill, 1994, 1997.)

Order	Description
Diptera	Ecologically diverse, the order is best known for containing species of medical and veterinary importance, such as mosquitoes and blackflies. However, there are also species that damage plants. These include the Psilidae (such as the carrot fly *Psila rosae*) and the Anthomyiidae (such as the cabbage root fly *Delia radicum*). Some species with carnivorous larvae (e.g. hoverflies) are used for biological control of crop pest insects.
Lepidoptera	The larvae of most species in this order feed on plants, and hence it contains many crop pest species. Some of the most important are members of the following families: Tortricidae, the larvae of which feed by boring into fruit, stems, buds or leaves; Pieridae, containing larval pests of legumes and crucifers; Noctuidae, the larvae feed on all above-ground parts of the plant. Noctuid pests are often referred to as 'worms' and they include cutworms (living in soil or in leaf litter and cutting stems of young plants), fruitworms, rootworms, bollworms (which feed on cotton), as well as species that burrow into stems (stem borers) and armyworms (foliage-feeding caterpillar species that occasionally form massive outbreak populations that move as a group to new areas of vegetation having exhausted their food supply).
Hymenoptera	This order contains many species of beneficial insects, including parasitoids of crop-feeding insect pests, which are used as biological control agents, and pollinators of crops, for example the western honeybee *Apis mellifera*. The larvae of some hymenopteran species in the suborder Symphyta (sawflies) are plant feeders, consuming foliage and tunnelling into tissue. They are important pests of forestry, particularly conifers. *Sirex noctilio* is a European species that has invaded Australia and caused significant damage to conifer forests. European spruce sawfly *Gilpinia hercinia* is native to Europe but is also an invasive species in North America where it is a major pest of forestry.

Continued

Table 2.1. *Continued*

Order	Description
Coleoptera	The more important crop pests can be found in the following families: Scarabaeidae, which include cockchafers and white grubs, whose larvae feed in plant roots and are important pests of trees and turf, while the adults feed on foliage; Elateridae, larval forms known commonly as wireworms that eat roots of cereals and bore into potato tubers; Meliodae, which include pollen beetles and longhorn beetles, which are important pests of trees; Chrysomelidae, leaf beetles; Curculionidae (weevils), one of the largest groups of beetles showing a lot of variation and containing many agricultural pests, the adults generally feed on foliage while larvae feed on stems, roots, seeds and fruits depending on the species.
Hemiptera	The Homoptera usually feed on phloem (plant sap) and some are vectors of viral plant pathogens. Serious pests occur within the following families: Delphacidae (plant hoppers), include species that feed on rice and sugarcane; Cicadellidae (leaf hoppers), include species feeding on crops such as rice; Aleyrodidae (whiteflies), important virus vectors and pests of glasshouse crops, citrus and cotton, include some serious polyphagous species such as the tobacco whitefly *Bemisia tabaci*; Aphididae (aphids), among the largest and most important groups of pests, aphids are highly fecund, reproducing during summer months by parthenogenetic vivipary (giving birth to live young without the need to mate with a male), and they vector plant viruses and cause direct feeding damage; Coccidae (scales and mealybugs), common in the tropics and subtropics where they are important pests of citrus and other tree fruits. Important crop pests in the Heteroptera are found in the families Miridae (capsid bugs), which feed on crops such as cotton, tea and cacao in the tropics, and the Scutelleridae (shield bugs).
Thysanoptera	In addition to causing direct feeding damage on plants, some thrips species are vectors of plant pathogenic viruses. One of the most serious crop pests is the western flower thrips, *Frankliniella occidentalis*. A polyphagous species, it feeds on vegetables, cotton, flowers and citrus in its home range in America. It is also an invasive species in Europe where it is highly damaging to glasshouse crops and is a vector of plant pathogenic viruses and fungi.
Orthoptera	Most important pests are located in the family Acrididae (short-horned grasshoppers and locusts) – about 500 species are considered to be important crop pests. Complexes of acridid species form the dominant insect grazers in the main grassland areas of the world; therefore they form important pests of agricultural grassland and cereals grown in these areas. In some species, massive outbreak populations can occur. These include the dispersive swarms of the desert locust, which can cause devastation to crops and other plants over wide areas. Locust swarms occur on one-third of the world's land surface and have been recorded as causing crop devastation since biblical times.

pathogens, e.g. Dutch elm disease. Some insects are important vectors of diseases of people and livestock. For example, mosquitoes are vectors of malaria, dengue virus and West Nile virus; bubonic plague is transmitted by the oriental rat flea *Xenopsylla cheopsis*; Chagas disease is vectored by the kissing bug *Triatom infestans*; human sleeping sickness and animal trypanosomiasis are carried by tsetse flies; and bluetongue, a serious virus disease of ruminants and cattle, is transmitted by the midge *Culicoides imicola*. Insect pests can also be particularly problematic if they have developed resistance to the chemical pesticides used against them, or if they move outside their native range into new territories. We must remember, however, that most insects are not pests. Many insect species provide vital, beneficial environmental services, for example as pollinators of crops and wild plants. There is also a wide array of predator and parasitoid species that consume crop pests and thus act as biological control agents.

Case study: the Colorado potato beetle
The Colorado potato beetle, *Leptinotarsa decemlineata* (Coleoptera: Chrysomelidae), is the most serious pest of cultivated potato and also one of the most difficult of all agricultural insect pests to control (Hare, 1990). It is thought to have originated in central Mexico, but was first described from a specimen feeding on buffalo bur, *Solanum rostratum*, collected from the US Rocky Mountains. The adult beetles are about 10 mm long, with a yellow-orange coloured body and black stripes. The beetle is confined to feeding on solanaceous plants and has been recorded feeding on about 20 species. In the 1860s it expanded its host range and started feeding on cultivated potato, *Solanum tuberosum*, in the USA. It quickly became a serious pest of potato throughout the eastern USA. It was accidentally introduced into Europe during World War I. It is now established throughout Europe with the exception of the British Isles. The adult beetles overwinter in the soil. In spring, females lay egg masses, comprising 20–60 eggs, on the underside of leaves. In laboratory experiments, females were recorded to lay up to 4000 eggs each. The larvae and the adults feed on plant foliage. If not controlled, then populations of the beetle will defoliate potato plantings before tubers are formed.

Because of its seriousness for potato production, Colorado potato beetle was the first agricultural pest to be controlled on a large scale with synthetic insecticides. Dichlorodiphenyltrichloroethane (DDT) was used very widely against it after World War II and was very successful initially. However, the beetle has a propensity to develop resistance rapidly to pesticides, which may be connected to its ability to tolerate the glycoalkaloids produced by solanaceous plants. Resistance to DDT evolved as early as the 1950s. It has subsequently developed resistance to all types of pesticide deployed against it including modern compounds such as neonicotinoids (Alyokhin *et al.*, 2007). Pesticide resistance appears to occur more rapidly each time a new active ingredient is used.

Acarina: mites
Mites (subclass Acarina) are members of the class Arachnida, which also includes spiders and scorpions (Lindquist, 1984; Walter *et al.*, 1996). Most species of the Acarina have no detrimental impact on humans. Indeed, many perform vital functions within natural and managed ecosystems as decomposers and predators. However, there are still a large number of acarine species that damage crops. A brief summary of the pest status of the different acarine orders is given in Table 2.2. Most acarine pests of crops are

Table 2.2. Acarine pests of agricultural, veterinary and medical importance. (Detailed reviews can be found in Helle and Sabelis, 1985a,b; Sonenshine, 1993; Hill, 1994; Lindquist et al., 1996.)

Superorder	Order	Pest status
Actinotrichida	Astigmata	Generally minor pests. Include pests of stored products such as the flour mite, *Acarus siro*. The bulb mite, *Rhyzoglyphus echinopus*, is a pest of bulbs, including onion, and mushrooms.
	Prostigmata	The Eriophyidae contain about 40 species that significantly affect crop production. Some of the more important species include: the apple rust mite, *Aculus schlechtendali*; the citrus bud mite, *Aceria sheldoni*; the citrus rust mite, *Phyllocoptruta oleivora*; the tomato russet mite, *Aculops lycopersici*; the coconut mite, *Aceria guerreronis*; and the wheat curl mite, *Aceria tulipae*.
		The Tetranychidae contain about 20 species of important crop pests. Important species include the two-spotted mite, *Tetranychus urticae*, and the green cassava mite, *Mononychellus tanajoa*.
Anactinotrichida	Ixodida	Over 100 species that are important pests of man and livestock. Transmit more blood-borne diseases than all other arthropods. Key genera include *Ixodes*, *Ripicephalus*, *Amblyomma*, *Dermacentor* and *Boophilus*.
	Mesostigmata	The Mesostigmata contain a small number of important pests. The red poultry mite, *Dermanyssus gallinae*, is a vector of viruses to domestic fowl. The varroa mite, *Varroa destructor*, is a major pest of honeybees.

contained within the Prostigmata, especially the Eriophyidae and the Tetranychidae, which are obligate plant feeders. Members of these groups attack a wide range of cultivated plants including fruits, vegetables, cereals, coconut, cassava, sugarcane and ornamentals (Hill, 1994). The Tetranychidae feed by piercing the epidermal tissue of the host plant and feeding from cells underneath this layer (Bondada et al., 1995; Warabieda et al., 1997). Once considered to be of secondary importance, they have become serious pests because of their capacity to develop resistance to pesticides (Cranham and Helle, 1985). In Africa, the exotic cassava green mite, *Mononychellus tanajoa*, introduced accidentally from South America, is a major threat to cassava production (Yaninek and Herren, 1988). The Eriophyidae are second in importance to the Tetranychidae as phytophagous pests, but still contain some economically important species. Of the approximately 3000 known species of eriophyoid mites, some 40 species are considered major pests (Lindquist et al., 1996). Some species transmit viruses. The wheat curl mite, *Aceria tulipae*, for example, is the vector of the highly destructive wheat streak mosaic rymovirus, which is widespread in the grain belts of the USA and Canada (Styer and Nault, 1996). Note also that some species of Acarina – most notably in the order Ixodida (ticks) – are important veterinary and medical pests, feeding on blood and transmitting diseases to their hosts. The order Mesostigmata also contains some pest species of animal hosts. The varroa mite, for instance, is a pest of European honeybees, having switched from its original host the Asian honeybee in the 20th century, and has had a highly detrimental impact on European honeybee populations in nearly all countries where they are kept.

Molluscs: slugs and snails
There are thought to be between 90,000 and 250,000 species of molluscs. The majority live in marine environments and include the cephalopods (octopus and squid), which have well-developed nervous systems and advanced functions. Terrestrial slugs and snails occur within the gastropod group of molluscs and are very important plant pests in some countries (Godan, 1983). Feeding occurs both above and below ground, on living plant tissue and on decaying vegetation. Small leaves can be completely consumed while large ones may have holes in them or be eaten around the edge. Most damage and plant loss occurs during plant establishment. Additional crop damage occurs as a result of the growth of bacteria and fungi on snail faeces and slime. In the temperate climate of central Europe snails are of less significance as pests than slugs. Important species include the European field slugs *Deroceras reticulatum* (Muller) and *Deroceras agreste* (L.). These species are highly polyphagous, attacking autumn-sown cereals and main crop potatoes, peas, clover, root crops and reseeded rye, especially in the seedling stages.

Plant pathogens

All species of crop plants are susceptible to infection by parasitic microorganisms, which together cause 12–13% loss of potential global crop yield, although yield loss varies widely between different crop species (Pimentel, 1997). Ten different groups of microorganisms have evolved to be parasitic on plants and cause significant levels of disease: fungi, algae, oomycetes, plasmodiophoromycetes, trypanosomatids, bacteria, phytoplasmas, nematodes, viruses and viroids. With such a wide array of different types of pathogen species it is not surprising that understanding plant pathogen biology is crucial to sustainable disease control. Plant pathogens of crops are widespread and common in all countries. In economic terms, the most important pathogens are fungi, oomycetes and viruses. Each crop species is exploited by an average of about 100 different plant pathogen species (Agrios, 2005). They infect all parts of the plant (roots, stems, leaves, vascular system) and all life stages from seed to seedling to mature plants, although generally it is the seedling stage that is most susceptible. A single pathogen species that can infect a range of host plant species can, as a result, be responsible for many different diseases. For example, it is estimated that North American crops suffer from 100,000 different diseases caused by 8000 different fungal species (Agrios, 2005).

Plant pathogens are broadly categorized by the mechanisms by which they obtain nutritional resources from the host plant. Necrotrophs use toxins and cell-wall-degrading enzymes to rapidly kill host plant cells and consume the plant cell nutrients that are released. In contrast, biotrophs require living plant cells in order to grow and complete their development (Carlile *et al.*, 2001). As a result they are often specialists and have evolved a complex set of genes and transcriptional responses associated with virulence as a result of the arms race for survival between pathogen and host. The negative impact of biotrophs on the host plant occurs as a result of the pathogen diverting host nutrients for its own purposes. Pathogens that are described as being hemibiotrophic use a combination of strategies. The initial phases of infection are marked by biotrophic growth on host tissues but then as infection proceeds there is a switch to necrotrophy with extensive cell death. Pure biotrophs are obligate parasites. However, a range of species of fungal and bacterial plant pathogens are able to grow both on living hosts and on decaying organic matter (i.e. saprotrophy) at different stages of their life cycle. Plant disease manifests itself as a result of a three-way interaction between the pathogen, the plant host and the environment (Strange, 2003).

Fungi
There are estimated to be about 1.5 million species of fungi. Most are saprotrophs (i.e. feeding on dead organic matter). Relatively few species are parasites of animals, but there are estimated to be over 10,000 species that are parasites of plants. In fact, all plant species are susceptible to fungal pathogens. Infection generally occurs by the germination of spores on the host surface, with subsequent development of hyphae on or in the plant. Life cycles of individual species vary widely and they can be complex. Some species are confined to growing on the plant epidermis, while some penetrate plant tissues between cells and others grow through plant cell walls or through the vascular system. Transmission can be by vegetative growth or in some cases by flagellated zoospores, which are able to swim a short distance in free water. However, most fungal plant pathogens are transmitted by non-motile spores. In particular, the production of large numbers of asexual spores appears to be a key feature of the rapid spread of fungal plant pathogens, and they can be transported long distances by air currents or vectored by animals. Some fungal species produce resting structures which remain inactive in soil for long periods until contact is made with a susceptible plant.

Five phyla of true fungi are recognized: *Chytridomycota*, *Zygomycota*, *Glomeromycota*, *Basidiomycota* and *Ascomycota*. Examples of some key fungal plant pathogens are given in Table 2.3. The largest number of plant pathogenic species is located in *Ascomycota*. Members of this phylum are characterized by the production of a sac-like structure (the ascus) that contains the sexual spores. Infection can occur from both sexual and asexual spores. There are many species in the *Ascomycotina* in which the sexual phase is not known and which reproduce entirely asexually, and which have only been confirmed to be members of the *Ascomycota* by using molecular (DNA) data. Some ascomycete plant pathogens produce toxins which are highly dangerous to mammals if ingested and which have historically caused severe chronic and acute problems for people and their livestock. For example, fungal species of the

Table 2.3. Some important plant pathogens occurring within the true fungi. (From Lucas, 1998; Agrios, 2005.)

Phylum	Class	Genus	Example of host plants/disease
Zygomycota	*Zygomycetes*	*Rhizopus*	Soft rots and postharvest decays of fruit and vegetables
		Mucor	Stored vegetables and fruits
		Olpidium	Infects brassicas and transmits pathogenic viruses
Ascomycota	*Dothideomycetes*	*Alternaria*	Foliar diseases, stem rot and fruit rot and postharvest disease on a wide range of fruit and vegetables
		Septoria	Leaf spot and blight on fruit, vegetables and cereals
	Eurotiomycetes	*Aspergillus*	Aflatoxins
	Leotiomycetes	*Erysiphe*	Powdery mildew on a range of ornamentals and vegetables
		Botrytis	Blossom blights and fruit rots in fruit, vegetables and ornamentals; postharvest decays of fruits and vegetables
		Sclerotinia	Rots and moulds in very wide range of vegetables, ornamentals and field crops
	Sordariomycetes	*Nectria*	Cankers in many tree species
		Magnaporthe	Rice blast
		Colletotrichum	Anthracnose diseases in many annual crops and ornamentals
		Claviceps	Ergot in cereals and grasses
		Fusarium	Wilts on vegetables, flowers, some plantation crops; head blight on grains, root and stem rots of vegetables, flowers and field crops; postharvest decay on vegetables and ornamentals; mycotoxins
		Ophiostoma	Dutch elm disease (wilt)
		Gaeumannomyces	Take-all disease of cereals
Basidiomycota	*Pucciniomycetes*	*Pucccinia*	Severe diseases on crops from the grass family – wheat, rye, barley, oats, sugarcane. Some species cause disease in cotton, vegetables or ornamentals
		Uromyces	Rusts on legumes
		Phakopsora	Soybean rust
	Ustilaginomycetes	*Entyloma*	Rice leaf smut

genus *Claviceps* are pathogens of cereals. They grow in developing cereal grains on the maturing plant and replace the seeds with fruiting structures known as ergots. These contain fungal alkaloids and when eaten they are highly toxic; effects include skin blisters, convulsions, hallucinations (known as St Anthony's fire), permanent mental damage, spontaneous abortion and gangrene. Ergotism has resulted in countless deaths, probably since the start of cereal cultivation. A range of mycotoxins are also produced by other fungal species growing on both live plants and harvested crops. Aflatoxins, produced by the fungus *Aspergillus flavus* and related species, are carcinogenic and highly toxic to people and livestock, and are produced on fungal-infected cereal grains, legumes and nuts. Chronic and acute effects occur depending on the dose received.

The *Basidiomycetes* contain the rusts and smuts, which are obligate biotrophic plant pathogens and contain a number of economically important species. Rusts are members of the class *Pucciniomycetes*. They are among the most destructive of plant diseases. These are obligate plant pathogens that infect leaves and stems. For example *Puccinia* species cause severe infections on wheat, other cereals and also crops such as cotton. Most species within the genus are highly specialized, infecting just a small number of plant species or genera. Smuts are members of the class *Ustilaginomycetes*. They have a worldwide distribution and comprise about 1200 species. The plant is not normally killed by the infection but growth can be severely restricted.

Oomycetes
Some of the most important plant diseases are caused by organisms that have very similar morphologies and life histories to fungi, but have been shown by molecular studies to be members of different biological taxa and thus evolved the ability to parasitize plants independently. Of these the most important are the oomycetes, which were considered to be true fungi until about 1990, but are actually part of a large group of eukaryotic protists termed the stramenopiles that includes the brown algae and diatoms. The oomycetes produce infective, motile haploid zoospores that are released from sporangia. They swim over the plant surface, germinate and grow into the plant tissue to cause disease. Sexual reproduction – which may or may not be present – leads to the production of diploid zoospores, which enable survival over winter.

Some of the most important oomycete plant pathogens are listed in Table 2.4. The main oomycete plant pathogens are contained within two orders – the *Saprolegniales* and the *Peronosporales*. The latter contains some of the most important of all plant pathogens. *Pythium* species cause damping off on seedlings, seed and root rots. *Phytophthora* species cause blights, fruit and root rots in a range of plants (and include *Phytophthora infestans*, the cause of late blight of potato, which resulted in the Irish famine). *Phytophthora ramorum* is an emerging disease that causes sudden oak death and has resulted in significant losses of oaks in California and Oregon; it has recently arrived in Europe. The oomycetes also include the downy mildews, which contain different groups that infect mono- and dicotyledonous plants, respectively.

Table 2.4. Some oomycete plant pathogens. (From Lucas, 1998; Agrios, 2005.)

Species	Example of host plants/disease
Pythium spp.	Damping-off diseases of seedlings of a very wide range of plants. Also cause seed, root and fruit rots
Phytophthora infestans	Potato late blight
Phytophthora ramorum	Sudden oak death
Peronospora spp.	Downy mildews in a range of plants, e.g. onion, tobacco, soybean, lucerne, clover
Hyaloperonospora brassicae	Downy mildew of brassicas
Plasmopara viticola	Downy mildew of grapes

Examples include *Bremia* and *Peronospora* species, which infect a range of fruit and vegetable hosts. *Hyaloperonospora brassicae* causes downy mildew on brassicas. *Albugo candida* causes white rust on crucifers.

Bacteria

There are about 1600 known bacterial species and only about 100 of these are parasites of plants. *Agrobacterium tumifaciens* causes the disease crown gall and infects about 200 different plant species. Species in the genus *Erwinia* cause blights, wilts and soft rots in many crop species. For example *Erwinia amylovora* causes the disease known as fireblight, a serious disease of apples and other species. *Xanthomonas campestris* pv. *campestris* is the main bacterial disease of crucifers and can cause severe epidemics in these crops.

Viruses

There are about 700 different viruses pathogenic to plants and they include some that are highly important economically (Strange, 2003). Viruses comprise nucleic acid (single-stranded or double-stranded DNA, or single-stranded RNA) in a protective protein coat. Most plant viruses are of the single-stranded RNA type. Host cells become infected when viral nucleic acid passes through the cell membrane and uses the host cell metabolism for its own reproduction. Infection occurs through wounds made mechanically, by a vector or by vertical transmission from infected pollen grains. Some of the most serious plant pathogenic viruses are vectored by insects and there is often a high level of specificity between virus and vector. For example, barley yellow dwarf virus – a member of the luteoviruses – is transmitted by aphids and infects 150 species of grasses including all cereals. The largest number of plant pathogenic viruses (over 100 species) is contained within the potyviruses (named after potato virus Y, PVY) and they affect crops such as sugarcane, beans, salad and root crops. Some of the most economically important potyvirus species are transmitted by aphids. Geminiviruses are single-stranded DNA viruses and cause serious diseases of vegetable crops, being transmitted by whiteflies and leaf hoppers. They include African cassava mosaic virus (which is estimated to result in a 50% loss of the annual potential yield of cassava in Africa), maize streak virus, and tomato yellow leaf curl virus.

Plant pathogenic nematodes
Nematodes (phylum Nematoda) are unsegmented worm-like animals. They are very diverse (about 80,000 species have been recorded) and include many parasitic species of animals and plants (about 15,000 species). Most plant pathogenic nematodes are associated with the soil. They reduce both crop yield and market value by causing cosmetic damage. Infection is generally done by juveniles, which hatch from eggs and penetrate young roots to feed on plant tissue. Important groups include root-knot nematodes, *Meloidogyne* spp., which infect over 2000 plant species including vegetables as well as staples such as potatoes and yam. Infection results in the formation of galls, within which the nematodes feed and which make the roots appear knotted. Cyst nematodes are common in temperate regions, and consist of species in the genera *Heterodera* and *Globodera*. Examples include soybean cyst nematode, *Heterodera glycines*, and potato cyst nematode, *Globodera pallida*.

Case study: the Irish Famine
Crop losses to some plant pathogens have resulted in severe famines that have had global geopolitical repercussions. Perhaps the best known example in the Western world is the Irish Famine of the mid-19th century. In the 1800s, cultivation of potato (*Solanum tuberosum*) became widespread in Europe because the crop gave excellent yields (Agrios, 2005). Potatoes were particularly well suited to the cool and wet climate in Ireland and gave good yields to feed poor tenant families who had to survive on very little land. By the 1840s a third of the population was entirely dependent on the crop. However, crop failures occurred in 1845 and 1846 as a result of a new disease, termed late blight, which had not been present in Ireland before and which caused up to 75% of the crop to be destroyed. The cause of the disease was not discovered until 1861 when a fungus (which is now known to be an oomycete pathogen), *Phytophthora infestans*, was shown to be the infective agent. The disease was accidentally introduced with potatoes from central Mexico, which is now known to be the centre of origin of the disease. It was first recorded outside Mexico in the USA in 1842 and appeared in Europe shortly thereafter. The blight epidemics of 1845 and 1846 occurred across Europe and in North America, but they were far more severe in Ireland than elsewhere because of the widespread habit of keeping potatoes in underground stores after harvest, which were also wiped out by blight. Cool moist weather that favoured potato production also favoured disease. The situation was inflamed by a range of socio-political and economic factors including restrictions on aid imposed by the British government. The failure of the potato harvest led to famine, collapse of civil society and associated epidemics of cholera, dysentery and other diseases. Family breakdown, starvation, disease and forced expulsion resulted in many leaving the land. The famine is thought to have resulted in one and a half million deaths while the same number of people emigrated, mainly to North America. The potato famine remains a source of much historical debate and retains a strong cultural resonance within the Irish Diaspora to this day.

Today, late blight is controlled using a combination of cultural practices, pesticide sprays and varietal resistance. The centre of origin of *P. infestans* in Mexico is also a centre of genetic diversity for *Solanum* species closely related to the cultivated potato, *S. tuberosum*. As a result of the co-evolution of *P. infestans* and its host species, resistance genes were discovered in native *Solanum* species (Grunwald and Flier, 2005). However, when single gene resistance was bred into cultivated potato it was rapidly overcome. The emphasis now is on breeding durable resistance based on quantitative traits (Solomon-Blackburn et al., 2007).

Phytophthora infestans can reproduce asexually and sexually. Sexual reproduction requires the presence of two mating types (A1 and A2) in the pathogen population. The late blight pathogen that was accidentally introduced into Europe in the 1800s consisted of just one genetic strain of the A1 mating type (Fry et al., 1992; Goodwin et al., 1994). The A2 mating type was accidentally introduced into Europe in the 1980s. This facilitated sexual recombination in the species and as a result new strains of the pathogen have evolved. Some of these have increased aggressiveness to host plants and resistance to metalaxl fungicides (Agrios, 2005).

Weeds

Weeds are undesirable plants. This definition is dependent on context, since a plant species that is undesirable in one situation may be desirable in another. For example, many arable weeds reduce the yield or quality of a crop but they also serve as a resource for beneficial invertebrates and birds, and thus are important for farmland biodiversity and conservation of wildlife. Weeds can be placed into four overlapping categories.

1. Agricultural weeds interfere with farm production by competing with crops for resources (soil nutrients, light, water), by contaminating the harvest with their seeds, by acting as reservoirs for invertebrate pests and plant pathogens, or because they are poisonous to people or livestock. There are estimated to be around 500 major species of weedy plants in the USA (Hajek, 2004) and in the UK there are about 300 potential weed species found in arable land alone (Naylor and Drummond, 2002). Approximately 40% of agricultural weeds occur in two families: the *Poaceae* (grasses) and the *Asteraceae* (composites and asters).
2. Invasive weeds are plant species that have been introduced to an area, region or country where they are not indigenous and which subsequently become established and spread.
3. There are also a number of weed species that are parasitic on other plants.
4. Finally, 'volunteers' are crop plants that grow in undesired locations. These can be a real problem when the farmer practises crop rotation, i.e. if crop plants cultivated last season survive over winter as seeds or plant debris and grow up among the following crop grown in same field. Some crops such as oilseed rape are also good at 'escaping' from the field and growing as weeds elsewhere on the farm.

In ecological terms, many agricultural weeds are ruderal species, i.e. plants that are adapted to colonize disturbed habitats. Weed establishment occurs from seeds that are carried by the wind, by animals, by human activity (e.g. on farm machinery) or by the germination of seeds that are already present in the soil (i.e. the weed seed bank) (Radosevich *et al.*, 2007). Cultivated fields are continually being disturbed, and this creates an unusual form of cyclical plant succession that favours the growth of ruderal weed species. The weed flora of agricultural fields tends to be dominated by herbaceous annual species that have a greater competitive ability in the seedling stage than perennial plants. Weeds are also favoured by features of agricultural production that counteract the normal limitations on plant population development. This can include an absence of herbivores and soil nutrient enrichment by application of fertilizers. While many agricultural weeds are opportunists, it would be wrong to think of them all as ecological generalists. Agrestals, for example, are specialist plants that require the disturbance associated with tillage and ploughing, and are associated with particular cropping systems.

Pest Control

Today, most farmers and growers are reliant on chemical pesticides for pest management. Alongside advances in mechanization, crop breeding, irrigation and synthetic fertilizers, they have been of tremendous benefit in increasing crop yields in the last half century. However, their use is becoming more difficult due to the evolution of resistance in pest populations, pest resurgence and the emergence of secondary pests following the destruction of natural enemies. An increasing number of pesticide product withdrawals are happening because of health and safety legislation or lack of profit for the product manufacturers, and this is adding to the challenge. All of these factors are reducing the availability of effective compounds. Further pressures on pesticide use are happening where consumers and other groups are expressing concerns about the safety of pesticide residues in food. When used injudiciously, synthetic pesticides may also cause harm to the environment. Some older compounds have caused serious health effects in agricultural workers, primarily because of inadequate controls and safety equipment during handling, and unsurprisingly this occurs nearly always in economically deprived countries.

Farmers and growers in many countries are trying to reduce the amounts of conventional chemical pesticides used, in response to demands from retailers, governments and pressure groups. This needs to be done without sacrificing crop quality and agricultural productivity. The latter is a key issue: the world population is forecast to reach 7 billion people in 2012 and to be greater than 9 billion by 2050 (UN, 2004, 2009). This means that production of food and other agricultural crops will have to increase significantly in future years. However, this must be done without putting additional strain on natural ecosystems and the global climate (agriculture is a major contributor to greenhouse gas emissions).

Historically, many of the improvements in crop protection were by-products of new developments in agronomic practice. The use of crop rotations, growing local crop varieties and the better control of water and drainage are good examples of this (Thacker, 2002). Rational inquiry into crop protection, driven by theory and experiment, is a relatively modern endeavour. Systematic advances in crop protection only really started during the latter part of the 19th century with the use of nicotine for insect control and the development of inorganic salts, such as elemental sulfur and Bordeaux mixture, against plant pathogenic fungi (Thacker, 2002). These compounds had low biochemical specificity and they tended to be toxic to mammals. They were superseded by the development of more specific, less harmful, synthetic organic pesticides during and after World War II. Synthetic organic pesticides revolutionized agriculture and their development marked the start of the true era of modern crop protection science. They were part of a set of related advances in technology that also included nitrogenous fertilizers, plant breeding, irrigation and sophisticated mechanization. Crop protection rapidly became reliant on technology and the intensive use of non-renewable resources. This approach has been dubbed 'industrial farming' and has become the dominant paradigm in the Global North. Alongside other components it is responsible for increasing agricultural production in Europe by 68% and in the USA by 100% since the 1960s (Pretty, 2008). However, industrial farming is not without significant costs. These include environmental pollution, soil erosion, pesticide poisonings of people, and a significant loss of the biodiversity upon which we depend for ecosystem services such as plant pollination, natural pest control, and clean air and water. We shall discuss these issues in more depth later, but it is tempting to conclude that, while society had sufficient knowledge to develop the agricultural technology to dramatically increase crop yields, it lacked the ability to use it in a sustainable way.

The development of the 'industrial' paradigm of pest management and the global adoption of chemical pesticides

The 30 or so years after the end of World War II saw a wide range of synthetic organic chemicals developed for use as pesticides. These compounds were effective, could be synthesized cheaply, and different application systems were available to suit agronomic practice, such as sprays, baits, dusts, drenches and fumigants. Their high efficacy, low cost and ease of application enabled reliable pest control for the first time. They were seen as a 'silver bullet' approach to pest management, one consequence being that they were used as prophylactics, with frequent, high-dose applications usually on a calendar basis irrespective of the size of the pest population. The aim of this approach was eradication of pest populations using chemical pesticides. As we shall see, it has resulted in significant negative effects.

The history of the development of these products has been characterized by improvements in the environmental performance of active compounds and a decrease in the amount of material applied to control a pest population.

Let us take insecticides as an example. Four major groups of synthetic organic insecticides were introduced widely from the 1940s: organochlorines, organophosphates, carbamates and pyrethroids. The first synthetic organic insecticide, the organochlorine (OC) DDT, became commercially available in 1942 (Thacker, 2002). DDT works by disrupting ion movement across neurons. It was cheap to produce, stable and persistent, selective for insects and with a broad spectrum of insect activity. It allowed high levels of control of many insect pests of agricultural crops for probably the first time in history and it has also been of major significance for the control of medically important insects. For example, it has been used widely to control mosquito vectors of malaria and it is still in use against mosquito pests today. However, the high persistence of DDT and other OCs means that they can bioaccumulate in the fat reserves of animals, becoming increasingly concentrated along the food chain and resulting in toxic effects in non-targets. Concerns about the mammalian toxicity of DDT and other OCs have resulted in these compounds being banned in much of the world. Development of OCs was followed rapidly by the introduction of organophosphates (OPs), which include common-name compounds such as parathion, fenitrothion and pirimiphos-methyl. OPs act as neurotoxins, with anticholinesterase activity, but they are less persistent in the environment than OC compounds. Some have systemic properties and thus can be applied to one part of a plant but still kill insects feeding on another part. This is highly valuable for insect pests that feed on the underside of leaves and are shielded from direct contact with pesticide sprays. OPs are often toxic to insects in the order Hymenoptera, which includes many species of parasitoid that are important natural enemies of insects, and hence this can create problems for use in IPM, which seeks to promote the action of natural enemies in the crop environment. Concerns about the ecotoxicological effects of OPs and their potential impact on people are leading increasingly to these products being withdrawn.

The two other major groups of insecticides from the post-war period were the carbamates and the pyrethroids. The carbamates (common names of carbamate molecules include aldicarb, carbofuran and bendiocarb) are closely related to the OPs, and like them have anticholinesterase activity. They were first introduced in the late 1950s. They tend to be broad-spectrum compounds that work by contact action or when the pest insect has eaten them. The pyrethroid pesticides are a class of neurotoxic compounds that work by affecting the activity of sodium channels in the plasma membranes of nerve cells. They are based on synthetic forms of pyrethrum, a natural extract of chrysanthemum with insecticidal properties, modified to improve their chemical stability. Pyrethroids have a very low toxicity to mammals and birds and thus are popular choices as insecticides. They are, however, highly soluble in water and toxic to fish and other aquatic life, and hence are not used in situations where they can contaminate watercourses, and they can also have negative impact on non-target insects as their insecticidal activity is not specific. Pyrethroids are used very widely against pests of agricultural and horticultural crops. Common names of pyrethroid molecules include deltamethrin, cypermethrin and fenfluthrin.

During the same period after World War II, new groups of synthetic chemical fungicides and herbicides were also developed. Synthetic fungicides included the dithiocarbamates, trichloromethylmercapto compounds (such as captan), imidazoles (inhibitors of ergosterol biosynthesis, an essential component of the cell membrane in true fungi, and which have systemic action), benzimidazoles (such as carbendazim and benomyl), acylalanines (compounds with systemic action that are effective against oomycetes) and carboxamides (used against *Basidiomycetes*) (Morton and Staub, 2008). Crude herbicides, such as oil wastes, thiocyanates and creosote, which had to be used in very high doses were replaced with selective compounds. The first selective, synthetic organic herbicide, 2 methyl-4,6-dinitrophenol, was developed in the 1930s–1940s and used as a selective herbicide on cereals. It has toxicity to mammals but is less toxic than the insecticides developed and used in this period; however, it was associated with causing health problems in operators when it was introduced as they were unaccustomed to safe operating procedures necessary for chemical sprays (Green *et al.*, 1987). Other selective herbicides developed in this period included compounds that mimic the plant growth hormone indol-3-acetic acid and cause abnormal plant growth, and are used as 'hormone' weedkillers in cereals. These include phenoxyacid compounds developed in World War II such as 2,4-D (2,4-dichlorophenoxyacetic acid), phenoxybutyrics and pyridine acid herbicides (e.g. picloram, a systemic compound used against broad-leaved weeds). Other herbicide compounds achieved selectivity though physical factors, such as absorption through roots rather than through foliage. These include herbicides such as paraquat, a foliage-acting herbicide. It is toxic to humans if taken as an overdose and the effects are irreversible, hence it has a notorious reputation as a poison as it used to be widely available to the public as a garden weedkiller. Glyphosate is a widely used organophosphorus compound (but not related to OP insecticides) that has very low mammalian toxicity and acts on plant foliage (Duke and Powles, 2008). There is also a wide variety of herbicide compounds based on the urea and triazine chemical families. Triazines are taken up by plant roots and inhibit photosynthesis, and include herbicides such as atrazine, which is used for weed control in maize (Knott, 2002).

Problems Associated with Using Pesticides According to the Industrial Approach to Farming

Viewed solely from the standpoint of increasing food production, industrial farming has been highly successful. Today most farmers and growers in the developed world rely on chemical pesticides for crop protection. Synthetic organic pesticides have made an enormous contribution to increases in crop yields and will continue to play a key role in crop protection. However, their use as part of industrial farming is not without significant external costs, as listed below. And while these pesticides will remain an important tool for farmers and growers, they are not a panacea for crop protection.

Broad-spectrum activity and environmental persistence
Until the 1980s, most of the pesticide compounds used tended to be broad-spectrum agents that were sprayed on a regular basis according to strict calendar dates or crop phenology. Until this period there was little attempt to target pesticide applications according to when pest populations were most active or at their most threatening (a process known as supervised spraying, supervised control or guided control). The injudicious use of broad-spectrum pesticides, applied without supervised control, in combination with other aspects of agricultural systems, can be a source of environmental harm and social tension (Tilman, 1999; Millennium Ecosystem Assessment, 2005). Concerns about the environmental and human safety impacts of pesticides were raised soon after the widespread adoption of the first generation of synthetic organic compounds following World War II and came to the fore with the publication of Rachel Carsons's book *Silent Spring* in 1962. As has been stated above, some pesticides can become bioaccumulated resulting in acute and chronic toxicity to animals at the top of the food chain. Ecosystem effects of pesticides can include disruption to microbial communities and their activity in soils. There can be other adverse effects on terrestrial and aquatic environments caused by pesticides killing non-target organisms. These can have multiplier effects by reducing the availability of food for species at higher trophic levels, for example by removing insects that act as food for birds. However, we must emphasize that because chemical pesticides vary significantly in their physicochemical properties, the extent of any negative impact varies according to the particular active ingredient.

Effects on non-target organisms
Basic economics means that pesticides with a broad spectrum of activity are more attractive to agrochemical companies and to farmers. Pesticides are expensive to develop and hence an active ingredient that kills a wide range of pest species will result in a larger market for manufacturers, while for farmers there are considerable savings to be made if only one product needs to be used for pest control. However, pesticides with too broad an activity will create problems by lowering the biodiversity of non-pest organisms. The negative effects of broad-spectrum insecticides on beneficial arthropods are well documented (Thacker, 2002). The pollination of crops by honeybees and other insects, for example, is thought to account for the production of about a third of our food plants, and thus actions that reduce pollinator numbers must be avoided (Pimentel, 1997). Conventionally, the effect of pesticides on non-target arthropods has been estimated by: (i) calculating the median lethal dose of the pesticide; and (ii) making an informed decision on whether this dose is likely to be received by non-targets in the field. However, it is becoming increasingly apparent that a range of sublethal effects can occur on beneficial arthropods and these need to be taken into account for a full understanding of the ecological impact of a candidate pesticide. These include physiological effects, for example on development, immune function, fecundity and sex ratio, as well as effects on arthropod behaviour such as navigation,

feeding and oviposition. These effects can interact, resulting in negative impacts on community ecology, and may result in economic losses due to reductions in pollination or natural enemy activity (Desneux et al., 2007). We must be careful in jumping to conclusions, however. The sublethal effects of pesticides are often temporary, and the risks can be managed by controlled application so that chemical residue does not come into contact with non-crop areas, such as hedgerows or field margins. Moreover, not all interactions are detrimental. For example, pyrethroid insecticides sprayed on to fields can interact additively with naturally occurring aphid parasitoids for aphid control (Desneux et al., 2007). This points to a need for improving our understanding of the effects of pesticides on non-targets that can be incorporated into the regulatory system and used to minimize negative effects.

Target pest resurgence
Target pest resurgence occurs when a pesticide kills both a pest and its natural enemies, and the pest population recovers rapidly afterwards because of the absence of natural enemies. Often the pest population will become larger than before the pesticide was applied. The size of the effect depends on the spectrum of activity of the pesticide and its persistence in the environment and will vary according to the active ingredient.

Secondary pests
Secondary pests can occur when previously innocuous organisms, which were maintained at low levels because of the action of natural enemies, increase to problem levels because a broad-spectrum pesticide has killed their natural enemies. This can lead to increased application of pesticides to control the new pest, which only serves to make the problem worse. For example, use of broad-spectrum pesticides for production of cotton in the USA from the 1960s to the 1980s led to an increase in the number of serious pest species from two to eight (Begon et al., 2006).

Pesticide resistance
Inappropriate use of pesticides can also lead to negative effects that reduce the effectiveness of the pesticide to control target pest populations. Heritable resistance can occur through decreased sensitivity of target cells or the evolution of mechanisms for pesticide detoxification. Individuals that have evolved genetically based pesticide resistance will be favoured by natural selection and their offspring will form an increasingly large proportion of the population over time while the selection pressure remains. The expression of the resistant phenotype will become apparent in the pest population as control failures. Generally, the risk of resistance developing increases if only pesticides with the same mode of action are used. This can lead to cross-resistance (i.e. resistance to different pesticides that have the same mode of action). Multiple resistances (resistance to pesticides with different modes of action) can also occur. Resistance to pesticides evolved soon after the first synthetic pesticides were introduced. For example, resistance to DDT was observed within 7 years of this pesticide first being used (Hajek,

2004). Insecticide resistance often evolves within 10 years and herbicide resistance within 10–25 years of introduction of a new compound (Palumbi, 2001). Worldwide, over 500 species of arthropod pests have resistance to one or more insecticides (Hajek, 2004). Over 347 biotypes of 200 species of herbicide-resistant weeds have been documented by the International Survey of Herbicide Resistant Weeds (WeedScience.com, 2010). Fungicide resistance occurs in only a few plant pathogens but it has been recognized as a problem for some while (Skylakakis, 1987; Milgroom et al., 1989; de Waard et al., 1993).

Pesticide resistance can be the driver for alternative control measures as part of IPM. A good illustration is the two-spotted spider mite, *Tetranychus urticae*, attacking glasshouse crops and many outdoor crops. Resistance to a wide range of pesticides in *T. urticae* populations has forced a change in its management in glasshouse crops from chemical to biologically based control using predatory phytoseiid mites (Hussey and Scopes, 1985; Coop and Croft, 1995). Elsewhere, such as on apples and citrus, control of tetranychids has moved increasingly to IPM management strategies that conserve natural enemies (McMurtry, 1985; Hardman et al., 1995; Croft and Slone, 1998; Lester et al., 1998).

Toxicity
Pesticides can have mammalian toxicity and, depending on the overall level of exposure, this can cause a variety of serious health effects in people. Human poisonings of agricultural workers and others exposed to pesticides can occur from handling pesticides during manufacture or application, or from environmental contamination, such as contaminated water sources. This is mainly a problem in the Global South in situations where those working with pesticides have inadequate access to protective equipment and proper training. Worldwide, 26 million cases of pesticide poisoning are estimated to occur annually, predominantly in developing countries, which also see the largest number of fatalities, estimated at 220,000 globally per annum (Paoletti and Pimentel, 2000).

Interactive effects
The problems associated with the industrial approach to pest management do not sit in isolation from other pest management concerns. For example, some species that have developed pesticide resistance are also invasive alien species, such as the varroa mite or western flower thrips. Different pest species also interact to create further problems. Thus – almost to add insult to injury – the pesticide-resistant, alien invasive, western flower thrips, *Frankliniella occidentalis*, is a vector of plant pathogenic viruses. The arrival of an invasive pest species can disrupt the current control methods used to control other pests. And we must remember that an organism that is a pest in one context may be beneficial in another, as we have seen already in the case of arable weeds, which are also important as food for farmland birds. There are wider issues too, for example the long-term viability of using pesticides that are

derived from fossil fuels. These issues make pest problems particularly complex to address and to solve.

Declining availability of products
Problems of resistance, resurgence and secondary pests associated with industrial farming can result in farmers entering a negative spiral in which they have to increase the number and type of pesticide applications year on year. Such an approach soon becomes untenable. Leaving aside the potential agronomic and environmental consequences of this approach, the rate at which new products are made available to farmers has declined rapidly since the 1970s. Due to a fall in the discovery of new active molecules and the increasing costs of registration, the availability of new pesticide active ingredients is declining (Thacker, 2002). New active ingredients are now estimated to cost around US$200 million and take about 10 years to develop. The availability of conventional, synthetic organic pesticide products is likely to decline further in the future. The number of agrochemical companies concerned with the development of new pesticides has reduced due to mergers, meaning that the crop protection market is now dominated by a small number of large corporations. The high costs of development will mean that these companies will concentrate on producing pesticides only for a small number of broad-acre crops such as cotton, rice, maize and wheat. At the same time, the pool of available products is being squeezed by increased government regulation or restrictions put on farmers by retailers. For example, at the time of writing, European Commission proposals for a new pesticides Regulation are being negotiated that will affect the whole of the EU. The Regulation will update the human and environmental safety 'cut off' criteria by which plant protection products are approved. This is likely to result in the prohibition of a significant number of synthetic chemical pesticide products. Impact assessments undertaken in the UK suggest that production of minor crops might be particularly affected, mainly because the majority of herbicides approved for these crops would no longer be approved. The loss of pendimethalin as a pre-emergence treatment would jeopardize weed control in cereals. The availability of fungicides that are considered to pose a hazard as endocrine disruptors could be reduced also, which might result in 20–30% yield losses in wheat due to an inability to control *Septoria tritici*, and would jeopardize disease control on oilseed rape leading to significant yield loss (PSD, 2008a,b).

Integrated Pest Management

New systems of crop protection must be effective in improving crop yield and quality, be safe for people, not cause unacceptable environmental harm, prevent the evolution of pesticide resistance and be adaptable to cope with emerging pests. In large part, the shortcomings of the industrial approach to crop protection stem from the fact that its practitioners do not utilize knowledge of agro-ecology to strategically control the use of pesticides. Most experts agree that the way to improve crop protection is through IPM. A pragmatic definition is that IPM is a systems approach that combines a wide array of crop protection practices with careful monitoring of pests and their natural enemies. IPM was defined formally in 1967 by the United Nations Food and Agriculture Organization Panel of Experts on Integrated Pest Control as:

> a pest management system that, in the context of the associated environment and the population dynamics of the pest species, utilizes all suitable techniques and methods in as compatible a manner as possible and maintains the pest populations at levels below those causing economic injury.
>
> (Bajwa and Kogan, 2002)

A more ecologically focused definition was given in Flint and van den Bosch's (1981) seminal book, *Introduction to Integrated Pest Management*:

> Integrated pest management (IPM) is an ecologically based pest control strategy that relies heavily on natural mortality factors such as natural enemies and weather and seeks out control tactics that disrupt these factors as little as possible. IPM uses pesticides, but only after systematic monitoring of pest populations and natural control factors indicates a need. Ideally, an integrated pest management program considers all available pest control actions, including no action, and evaluates the potential interaction among various control tactics, cultural practices, weather, other pests, and the crop to be protected.

IPM practices include resistant varieties, physical and cultural (i.e. cultivation) methods, biological controls and judicious use of modern, selective pesticides. These different tactics can be combined in various ways to suit local needs. Decisions can be made about what tactics to use, and when and where to use them, based on an understanding of the ecology of the pest, its natural enemies, and agronomic and environmental conditions. The aim of IPM is not pest eradication; rather it is the more realistic goal of reducing a pest population below its economic injury level.

In industrialized economies, IPM is seen as technologically based and is focused on using a suite of complementary control options in combination with pest monitoring and economic action thresholds. Here, IPM is a natural progression from supervised or guided control, in which the farmer uses economic thresholds to decide when to apply chemical pesticides. It is doubtful whether farmers and growers could take up IPM successfully unless they were already familiar with supervised control. In some developing nations, however, a different IPM model has been developed, based on training farmers to better understand the importance of natural biological control and to rely on their own observations in order to decide when to spray pesticides (Waage, 1997; Dent, 2000). In both situations, the goal is the same; namely to achieve a flexible and durable system that minimizes impacts on other components of the agro-ecosystem (Kogan, 1998). Examples of IPM practice in different crop types are given below.

IPM is significantly more complex compared with the simple 'spray-and-pray' approach to pesticides used in industrial farming. Therefore IPM programmes are best developed incrementally. There are many different types of

LEVEL ONE IPM

Multiple management tactics integrated for individual pests in each class of pests

↓

Multiple management tactics integrated for a single group of pests (arthropods, pathogens, weeds or vertebrates)

LEVEL TWO IPM

Multiple management tactics integrated across all groups of pests

LEVEL THREE IPM

All pest management tactics integrated with all farming practices

LEVEL FOUR IPM

IPM integrated with social, cultural and political realm; involve farmers, researchers, extension, industry, environmentalists and regulatory agencies

Fig. 2.1. Prokopy's four levels of integrated pest management (IPM). (From Prokopy, 1993.)

IPM programme and a wide range of levels of sophistication. IPM is also dynamic, in the sense that the pest problem is always changing; for example, when new pests emerge, when pesticide resistance develops or pesticides are withdrawn, when new crop varieties become available or when agricultural practices change. Prokopy (1993) set out four levels of IPM that are divided up according to their degree of sophistication (Fig. 2.1). The baseline level, Level One, concerns integrating different chemical, biological and cultural controls for the management of a single species of pest on a single type of crop. The top level, Level Four, involves incorporating all pest management practices within an overall Integrated Crop Management system that involves all members of the policy network (farmers, extension services, industry, retailers, regulators, non-governmental organizations, etc.) and takes account of the social, cultural and political context of farming. All four levels require sound information on: (i) the causes of the pest outbreak; (ii) the natural factors that limit outbreaks and which can be exploited for pest management; (iii) how the pest population changes over time and when pest control tactics should be deployed; and (iv) the costs and benefits of the different control tactics at the farmer's disposal.

Understanding the Natural Factors that Determine Pest Population Levels: Some Basic Ecological Concepts

IPM begins with understanding the life history strategy of a particular pest, together with the factors that naturally determine its abundance. Some inherent features of agricultural production cause pests to proliferate, but on the other hand there are also factors that limit the size and range of pest populations. An ecological approach enables a better understanding of why pest outbreaks occur and hence the effective measures that can be put in place to prevent or manage such outbreaks in ways that cause minimum disturbance to other components of the agro-ecosystem. Knowledge of pest ecology allows flexible, responsive pest management. For example, understanding the causes of pest outbreaks can allow changes to be introduced to farming practice that may prevent an outbreak occurring in the first place, such as encouraging populations of natural enemies or growing mixtures of crop varieties that slow down the spread of a pest population. It also allows predictions to be made of when pests are likely to be most active, and enables therapeutic agents to be applied only when necessary, as opposed to applying them regardless of pest numbers, as occurs with calendar-based spraying of pesticides. This not only saves the farmer money but also helps prevent unwanted effects such as the evolution of resistance to pesticides.

The factors that determine pest population dynamics interact in complex ways. This can make it difficult to obtain definitive evidence for individual mechanisms from experiments or observational data. We also need to distinguish between factors that influence the general abundance of a pest in farming regions and those that are temporally and spatially explicit, i.e. which cause a fluctuation in pest population levels during the course of a cropping season or within a single field. These factors include environmental conditions, intrinsic features of pest life history, intra-specific competition and interaction with other species, particularly natural enemies. It is the combination of factors affecting both the general level of abundance of a pest and its seasonal population dynamics that ultimately determine whether and when a pest population passes the point at which it becomes economically prudent to introduce control measures.

Life history strategies

The life histories and reproductive strategies of organisms have evolved by natural selection in response to their physicochemical and biological environment. They are encapsulated in ecological theory by the concept of r and K selection (MacArthur and Wilson, 1967; Statzner et al., 2001). In nature, there is a continuum of strategies and r- and K-selecting strategies lie at either extreme. The term r refers to the intrinsic rate of increase of an organism while K refers to the carrying capacity of a population that is limited by competition. Habitats can be r- or K-selecting.

Theory predicts that an r-selected population lives in an unstable, ephemeral or fluctuating habitat that is free from competition, for example a weed species that is adapted to colonize areas of bare ground. An r-selected individual is one that reproduces rapidly and produces many offspring, with little investment in each progeny. In contrast a K-selected individual has a low reproductive rate but puts more investment in progeny (e.g. by producing fewer but larger offspring and by protecting them from natural enemies). A

K-selected population lives in an environmentally stable habitat and will be a good competitor. It may be able to survive, for example, in a nutrient-poor environment (rust fungi that tolerate the harsh conditions on the phylloplane, with little water and lack of nutrition, are K-selected organisms) or with high levels of predation and parasitism, whereas an r-selected population will occur in an unstable habitat and is a poor competitor.

This concept provides a valuable way of thinking about how adaptation and natural selection results in particular patterns of birth, reproduction and death and is thus highly relevant in ecologically based pest management. Where an organism sits on the r- and K-selection continuum depends on the conditions of its environment and competition with other individuals. Plant pathogens such as *Rhizoctonia* and *Pythium* that attack seedlings or weakened plants are r-selected, for example. Any biocides applied against them will have to be fast-acting to cope with a rapidly developing pest population. Alternatively, a control measure could be put in place that limits the development of an r-selected pest population by providing it with high levels of competition. Such a strategy is used for the control of some plant pathogens by using antagonistic microbes that compete for soil resources. Some perennial crops such as fruit orchards can attract K-selected pests. In these circumstances ecological theory predicts slower-acting therapeutic agents could be used, but that trying to prevent pest build-up by introducing competitors is unlikely to be successful.

Factors that naturally limit pest populations

Physicochemical conditions
The physicochemical conditions of an environment help determine the occurrence and abundance of all living things including the properties of soil or water (pH, salinity), climate (temperature, rainfall, radiation) and weather. Most species of agricultural pests are poikilothermic (cold blooded) and temperature has a major impact on where they live and the rate at which their populations develop during the year. The effect of temperature on the development of poikilothermic organisms follows a defined mathematical relationship that can be used not only to determine the minimum, optimum and maximum temperatures for growth but also to identify how much thermal energy is required to reach a particular developmental stage. This information can be used to forecast times of peak pest activity and to target interventions such as pesticide sprays or the application of crop covers.

Competition
Individuals of the same species have very similar resource requirements, such as radiation (for photosynthesis), water, carbon dioxide, oxygen, mineral and organic nutrients, space or breeding partners. The supply of resources in an environment is often limited, and hence individuals from the same species will compete for them. This leads to intra-specific competition maintaining the population density within certain limits. Competition also occurs

between species. Ecological theory predicts that inter-specific competition can result either in the exclusion of one species by another or in stable coexistence, dependent on the amount of niche differentiation between species (Begon et al., 2006). This knowledge can lead to ecological solutions for pest management, for example by adjusting crop planting density to outcompete weeds or by the development of mathematical models that identify the best time to remove weed plants from a field.

Predation
As primary producers, crop plants are consumed by a wide range of herbivorous organisms including insects, mites, molluscs, nematodes and microorganisms. In turn, herbivores are consumed by carnivorous and parasitic organisms. The natural enemies of crop herbivores form a large part of the world's biodiversity and perform a vital function in limiting the populations of crop pests. It has been estimated that each agricultural pest species is fed upon by 50–250 natural enemy species. This vast number represents a highly valuable but underexploited resource for crop protection through biological control.

Factors that cause an increase in pest population size

Resource concentration
Crops are usually grown as monocultures. This presents potential pests with a concentrated resource, enabling them to increase their general level of abundance before intra-specific competition regulates pest population density. Evidence for the resource concentration hypothesis comes from experiments showing that phytophagous insects occur at denser populations in pure stands or larger patch sizes of plants (Root, 1973; Rhainds and English-Loeb, 2003) and that specialist species of phytophagous insects are less efficient at selecting a suitable host plant when it is growing in a diverse background of other, non-suitable species (Finch and Collier, 2000). Introducing a new crop species to an area will have similar effects in that it provides a new, concentrated resource for species able to exploit it. Genetically uniform crops, in which plant defences have been inadvertently removed through breeding, will be more susceptible to pests than populations of wild relatives. Ecological theory on the 'apparency' of different plant types to herbivores in natural ecosystems (Feeny, 1976) proposes that ephemeral plants are protected to an extent because they grow unpredictably in both time and space and thus are harder to locate. Therefore they are less likely to have evolved physical and chemical defences. This suggests that plant species that are exploited by humans as annual crops are more likely to be susceptible to consumers than perennial species.

Agricultural disturbance
Many broad-acre crops – rice, wheat, maize, cotton, soya – are planted once in a year in a particular field. Other annual crops such as field vegetables,

which are eaten fresh, are planted in sequential blocks within a field system in order to provide continuity of supply throughout the season. Both agronomic situations create an ephemeral and ecologically unstable environment. For example, after the ground has been made bare after ploughing, it is vulnerable to colonization by r-selected weed species that grow from the weed seed bank in the soil. A colonizing weed species that is able to grow in the newly disturbed habitat and has a high rate of reproduction can quickly build up to pest levels.

The perturbations associated with agricultural production also have a destabilizing effect on the relationships between pests and their natural enemies. Inappropriate use of broad-spectrum chemical pesticides can lead to resurgence in a pest population when the same pesticide spray kills natural enemies. The effects may vary depending on whether the natural enemies are generalists or specialist on one particular prey type. Generalist natural enemies are effective at preventing population outbreaks when the prey population is at a low level, whereas specialist natural enemies tend to dominate the regulation at high prey densities (Hesketh *et al.*, 2009). The population of a specialist natural enemy is dependent on that of a particular prey species, leading to a lag between the reproduction of the prey and the natural enemy that creates a cyclical or 'boom-and-bust' population pattern for both. Ecological theory indicates that stochastic effects associated with abiotic factors (e.g. changes in weather or other forms of environmental disturbance) can shift a pest population from a stable, low level regulated by generalist natural enemies to a cyclical, dynamic state characteristic of specialist natural enemies (Dwyer *et al.*, 2004). Needless to say, farmers want to avoid cyclical patterns of abundance of crop pests when the 'boom' part of the pest cycle results in damage to the crop above the economic threshold.

Changing physicochemical conditions
Agronomic practices designed to make the environment more favourable for crop production can also make it more favourable for pests. This could include improved irrigation and fertilization, which encourages the development of weeds, or growing crops under protection with raised temperatures, which encourages the development of invertebrate pests and plant pathogens. Stressing the crop, e.g. by drought, can also make it more vulnerable to pests. Seedlings are less able than mature plants to withstand detrimental changes to the physicochemical environment and as such are more vulnerable to damping-off diseases, etc.

Alien species
Some of the most serious pest problems arise through alien (i.e. non-native or non-indigenous) species being introduced accidentally to a new country or continent. About 10% of alien species establish and spread from their original point of introduction, and of these approximately 10% cause significant economic or ecological damage. More than 11,000 alien species have been documented in Europe alone, for example (DAISIE, 2009). Economic losses to crops from alien invertebrates, pathogens and weeds are estimated at

€4.5 billion per annum in the UK (approximately €75 per person), €55 billion for the USA, €4.5 billion for Australia and €72 billion for India (Pimentel *et al.*, 2001). Wheat yellow rust fungus *Puccinia striiformis*, for example, was accidentally imported into Australia in the late 1970s and had a major impact on wheat production as it established and spread through the country. Serious losses to European crop production have occurred since the early 1990s from alien insect pests, such as the western flower thrips *F. occidentalis* and the silver leaf whitefly *Bemisia tabaci* (Table 2.5).

The seriousness of alien invasive species can be illustrated by looking in more detail at a single example. In North America, the western corn rootworm (WCR), *Diabrotica virgifera* (Coleoptera), is a major pest of maize (Gray *et al.*, 2009). The larvae feed on roots of maize and some other grass species and the adults feed on the foliage and pollen. The pest originates in Guatemala, where it has been known for around 5000 years, and is thought to have spread and become a serious problem with the advent of intensive maize monoculture systems introduced by Spanish colonizers. It was resident in the western part of the US Great Plains in the 1870s and has since spread east to occupy about half the land mass of the USA with a rapid range expansion after World War II as maize production increased. WCR used to be controlled successfully by crop rotation with soybean, although insecticide treatments were still required and over US$1 billion was spent annually on its control in the USA. In 1995 a population in Illinois evolved that had lost fidelity to maize and was able to survive on soybean (Gray *et al.*, 2009). This new variant has spread within the USA. Farmers were then forced to switch their management practice towards use of insecticides and transgenic Bt maize. WCR has

Table 2.5. Alien invasive insect pest of importance in Europe. (From DAISIE, 2009.)

Species	Common name	Threat
Aphis gossypii (Hemiptera)	Melon and cotton aphid	Polyphagous pest of vegetable crops and citrus
Bemisia tabaci (Hemiptera)	Silver leaf whitefly	Polyphagous, serious pest of edible horticultural crops and ornamentals
Ceratitis capitata (Diptera)	Medfly	Larvae develop in wide range of fruits
Diabrotica virgifera (Coleoptera)	Western corn rootworm	Larvae feed on roots of maize
Frankliniella occidentalis (Thysanoptera)	Western flower thrips	Pest of outdoor and protected horticultural edible crops and ornamentals
Leptinotarsa decemlineata (Coleoptera)	Colorado potato beetle	Pest of potato crops
Liriomyza huidobrensis (Diptera)	South American leaf miner	Larvae feed on wide range of hosts especially vegetables and ornamental plants
Spodoptera littoralis (Lepidoptera)	Cotton leaf worm	Larvae feed on cotton, fruits and vegetable and ornamental crops

developed resistance to certain insecticides, making Bt maize the most viable option. WCR was accidentally imported to Belgrade airport in 1992, probably as a stowaway, and established in local maize fields. Since then it has spread to 20 European countries. Currently it is controlled in Europe using rotation with a non-host crop. However, it is now known that multiple introductions have occurred. This increases the risk that an introduction of the rotation-resistant variant will occur in Europe. This could create a dilemma because the current optimal strategy for WCR control is Bt maize, but there is resistance from the public to growing genetically modified (GM) crops in Europe. Research is being done on biological control, which is considered to be a potential market for organic maize and maize seed production fields. Elsewhere farmers rely on insecticides. However, there is a concern that farmers in the USA and Europe have largely abandoned basic IPM practices for maize and that this is likely to create problems for the future, such as insecticide resistance (Gray et al., 2009).

Alien pests can also damage native habitats by excluding other species, causing a loss of habitat for animals and microbes, altering plant community succession and interfering with the normal interactions of plants and other organisms, such as pollinating insects. Accidental introductions of some alien pest species have resulted in wholesale change in plant communities. For example, invasion of rangeland areas of the Midwest and western USA by cheatgrass, an annual grass, has resulted in the loss of much of the native sagebrush–bunchgrass range. In eastern North America, invasion by three insect species – the hemlock woolly adelgid (*Adelges tsugae*), the gypsy moth (*Lymantria dispar*) and the emerald ash borer (*Agrilus planipennis*) – has caused significant damage to native forests over millions of hectares. These invasions are proving difficult to control. For example, the hemlock woolly adelgid is expected to lead to the extinction of hemlock within a decade in some areas such as the southern Appalachians. For all three of these forest pests, invasion has resulted not only in major tree decline, but also significant effects at the community and ecosystem levels.

The success of alien invaders has often been explained in terms of escape from the natural enemies that would regulate them in their native range through inter-specific competition. However, formal support for the hypothesis was thin until two recent meta-studies of the scientific literature. Torchin et al. (2003) compared the parasites of exotic animal species in their native and introduced ranges. They found that the number of parasite species affecting native populations was twice that of exotic populations and the proportion of infected individuals was lower in introduced versus native populations. Mitchell and Power (2003) examined fungal, oomycete and viral pathogens of exotic plants in their native and introduced ranges. They showed that exotic populations had 84% fewer fungal and oomycete pathogen species and 24% fewer viral species than in their native range. Both studies lend clear support to the enemy release hypothesis. However, introduced species do not escape regulation by inter-specific competition entirely. In their analysis Mitchell and Power (2003) also found support for the biotic resistance hypothesis, which proposes that the invasiveness and impact of an alien species can be limited by competition with native species including natural enemies. A negative relationship was found between the noxiousness of exotic agricultural weeds and the number of pathogen species they had accumulated in the new range. Thus acquiring natural enemies in a new range can impact on invasiveness.

The natural enemies escape hypothesis leads to the obvious conclusion that introducing natural enemies from their original home range to the new area can counteract alien, invasive pests. This is a form of biological control known as classical control. However, escape from natural enemies is not the only explanation for the success of invasive species. The probability of successful invasion depends on a complex interaction between the characteristics of the habitat in which the alien species is introduced as well as the biological traits of the alien itself. These interactions vary on a case-by-case basis, although some general factors that contribute to the probability of invasion success other than natural enemy escape include: (i) the evolutionary history of the resident community (i.e. the idea that communities with intense competition will be resistant to invasions); (ii) community structure; (iii) the level of natural disturbance; and (iv) human activity such as pollution, local extinction of native species and accidental dispersal of alien organisms on vehicles (Radosevich *et al.*, 2007).

Emerging pests
Pests new to production agriculture can arise through the evolution of new pest strains. Increased reliance on a small number of pesticides and the widespread growth of crop species with limited genetic diversity are increasing the selective pressure for new pest strains to evolve. New strains can also evolve to overcome plant defence mechanisms. A new strain of the wheat stem rust fungus *Puccinia graminis* f.sp. *tritici*, termed Ug99, evolved in Uganda in 1999. It is now spreading towards Asia and Europe. It is able to overcome the resistance gene bred into standard wheat lines and as a result it is highly virulent, capable of causing 100% yield loss (Mackenzie, 2007). The potato blight oomycete *P. infestans* is a pathogen that can reproduce asexually and sexually. Sexual reproduction requires the presence of two mating types in the oomycete population. The pathogen that was accidentally introduced into Europe in the 1800s from its original home in Mexico consisted of just one genetic strain, i.e. a single mating type (Goodwin *et al.*, 1994). Unfortunately, the second mating type was introduced into Europe in 1976, facilitating sexual recombination in the species and as a result new strains of the pathogen have evolved. Some of these have increased aggressiveness to host plants and there are concerns that pesticide resistance may develop more rapidly.

Economic Thresholds

Understanding the factors that limit pest populations or which cause them to proliferate can inform us about suitable management approaches. This needs to be combined with economic thresholds to decide when to apply a pest treatment. It is self-evident that the financial costs of controlling a pest must be less than the benefit obtained in terms of improved crop yield or quality. The aim is rarely to eradicate a pest population; rather it is to manage it so that the population is maintained below the economic injury level. This is formally described as the pest population density at which the cost of pest damage equals the cost of the available control measures. On paper, the economic injury level represents the point at which it becomes financially viable to use pest management methods. It is normally thought of in terms of remedial treatments for pests (pesticide sprays, for example), but as a concept it applies equally to preventive treatments. In practice, once a pest population has reached the economic injury level it is usually already too late to apply the control measure, because it takes time for a control programme to work. The key decision metric in a pest management programme, therefore, is the economic threshold, which is the density of the pest where control measures must be exerted in order to prevent it reaching the economic injury level. The economic threshold is determined in essence by a cost–benefit analysis (Mumford and Norton, 1984). Economic thresholds have been important in the implementation of effective supervised pesticide spray programmes and they should be equally valuable in IPM. However, because they rely on detailed monitoring of pest populations and mathematical models of pest population dynamics they have been criticized, reasonably, for being too complicated and unwieldy for many farmers to implement (Orr, 2003; Rodriguez and Niemeyer, 2005). This criticism was made mainly in relation to resource-poor farmers in developing nations, but it is true to say that many farmers in the Global North also do not use formal, research-derived economic thresholds, although nearly all farmers scout their crops for evidence of pests and act accordingly. The most sophisticated economic threshold systems are used in high-value, labour-intensive crops that are subject to intense pest pressure and are heavily dependent on external inputs. In the Global North these would typically be greenhouse crops such as glasshouse salads, ornamentals and some soft fruit such as tunnel-grown strawberries, while in the Global South this would usually be horticultural edible and flower crops grown by large commercial operations for the export market.

Crop Protection Methods Used in Integrated Pest Management

In the past, under the industrial farming paradigm, the standard way of managing pests was by the application of chemical pesticides with little regard to the reason why the pest was present, the amount of crop injury being caused or the costs to the environment. Chemical pesticides were cheap and effective enough to make this an economically viable strategy. This is clearly no longer the case. From the information presented so far in this book, it should be apparent that the strategy used to control a particular pest should have a firm evidence base, i.e. it should be tailored according to the cause of the pest outbreak. Synthetic chemical pesticides have a vital role to play in modern crop protection, but they should be integrated with other methods in order to reduce the onset of resistance and, of course, to reduce harm to the environment and to people.

Although the crop protection industry is dominated in market size terms by conventional synthetic pesticides, there are a wide range of alternative methods that are being used increasingly by farmers and growers (in the context of this book, we use 'alternative' in the sense of giving farmers and growers a choice of different crop protection methods). These methods are commonly referred to in the literature under the following categories.

Modern selective pesticides
New knowledge on pest biochemistry, coupled with approaches such as molecular engineering and targeted screening for active compounds, have produced new synthetic pesticide compounds with far better environmental and human safety profiles than the 'old' chemistry of the 1950s and 1960s. Good examples of modern chemistry include synthetic insect growth regulators such as buprofezin and pyriproxyfen, which have high levels of selectivity, low mammalian toxicity and are classed by regulators as low-risk compounds (Grafton-Cardwell *et al.*, 2006). These new compounds are very important for crop protection (for a good review see Smith *et al.*, 2008). The development of new actives is a complex process involving the screening and characterization of a wide array of candidate molecules. Increasingly, these pesticide development programmes target specific sites within the cells and organelles of the pest organism. A huge array of bioactive molecules is synthesized in nature, and hence chemists are turning more to natural products as the leads for pesticide screening. In addition, the whole genomes of pest species are starting to be sequenced and this is paving the way for identifying new targets through comparative genomics and transcriptomics. However, as we have described previously in this book, the development of new synthetic pesticides is slow and expensive, and the benefits of a new compound may be lost quickly if pesticide resistance evolves. Thus, it is imperative to use new chemistry within an IPM framework that incorporates resistance management. Unfortunately, as we shall see shortly, most farmers still do not practise IPM and this is undoubtedly wasting the opportunity to get the full benefits from modern pesticides.

Cultural practices
Cultural methods of pest control involve changing the way a crop is grown in order to reduce or avoid pest damage. Crop rotation is one of the oldest strategies for managing pests and is particularly useful for controlling pest species with limited dispersal ability and host range (Minnis *et al.*, 2002). Rotation with non-host crops can break a plant pathogen's life cycle. Crop rotation may prevent the build-up of weed species that are adapted to any single crop or cropping system (Bond and Grundy, 2001). Other cultural control methods involve good crop management, such as destruction of crop residues, and careful choice of planting date to avoid key periods in pest life cycles, such as the laying of insect eggs (Ellis *et al.*, 1987). Techniques such as

intercropping, undersowing and companion planting may also reduce colonization by certain insects, pathogens and weeds (Theunissen and Schelling, 1996; Finch and Collier, 2000; Bond and Grundy, 2001).

Physical methods
Physical control methods are non-chemical/non-biological measures that destroy the pest, disrupt its development or activity, or modify the environment to a degree that is unacceptable or unbearable to the pest (Vincent et al., 2003). Examples include the use of mechanical weeders or growing crops under covers or with mulches. Mechanical in-crop weeding is possible in crops with wide row spacings (e.g. horticultural crops but not arable crops) and developments continue to be made in vision guidance systems for greater precision and intra-row weed control (Bond and Grundy, 2001).

Natural compounds
Living organisms produce a wide variety of molecules that can be used in pest management, such as biocidal compounds extracted from plants (often referred to as 'botanicals') or insect or plant signalling molecules (semiochemicals). Some plant compounds, such as pyrethrum, have been used as 'natural pesticides' for many years although compared with synthetic pesticides they are relatively unstable, have lower potency and greater price. Some new, highly effective pesticides are based on natural products from microorganisms, such as the insecticide Spinosad. Insect pheromones are used widely to monitor insect populations and are being used increasingly to control pest populations by disrupting the ability of insects to find mates.

Plant breeding
New crop cultivars with total or partial resistance can be bred using conventional methods or with genetic modification technology. Plant breeding can enhance plant resistance to pests, and considerable efforts are expended each year to develop new resistant varieties. There is some effort to breed for resistance against invertebrate pests, but most of the work has been done with developing varietal resistance against pathogens. Many hundreds of pathogen resistance genes have been identified in crop species. Unfortunately, most resistance conferred by single plant genes does not remain available for long because of the ability of many pathogens to overcome plant resistance through natural selection (Pink et al., 2008). Pathogen resistance that is determined by several plant genes tends to be more durable than single gene resistance. However, it is more difficult to handle in a breeding programme and is often partial, i.e. some disease still develops. The technical difficulties of dealing with multi-gene resistance are now being overcome with new technology such as marker-assisted breeding (Collard and Mackill, 2008). Partial resistance may not be an issue if some yield reduction is acceptable. However, in fruit and vegetables, where marketable yield is determined by quality, any disease blemishes may be unacceptable. Therefore it is vital that partial

resistance is used as part of an IPM programme to reduce pest damage to acceptable levels.

Biological control agents
For the purposes of this book, biological control is 'the use of living organisms to suppress the population of a specific pest organism, making it less abundant or less damaging than it would otherwise be' (Eilenberg *et al.*, 2001). The organisms used as biological control agents include predatory insects and mites, parasitoids, parasites and microbial pathogens and antagonists (Bellows and Fisher, 1999; Bale *et al.*, 2008). In Europe and North America, typical examples of where these agents are being used include: (i) the application of parasitoids to control whiteflies in glasshouses; (ii) parasitic nematodes against slugs; (iii) mycoparasitic fungi to control plant diseases of horticultural crops; (iv) use of viruses to control codling moth in apple orchards; and (v) building habitats on farms to increase natural populations of predators and other beneficial organisms. The ways in which biological control agents are used vary according to the type of pest (plant, microorganism, vertebrate or invertebrate), the biological characteristics of the control agent, as well as the agricultural setting. There are three broad biocontrol strategies: (i) introduction (release of an alien control agent to control an alien pest); (ii) augmentation (application of natural enemies that already live in the area of use); and (iii) conservation (manipulating agricultural practices or the environment to enhance natural control).

Biopesticides
There is no universally recognized definition, but in general terms biopesticides are mass-produced, biologically based agents manufactured from living microorganisms or natural products that are sold for the control of plant pests (Copping and Menn, 2000). In some countries, such as the USA, biopesticides also include GM plants expressing introduced genes that confer protection against pests or diseases (so-called plant incorporated protectants). Some authors consider that mass-produced 'macro' biological control agents (predatory insects, parasitoids, nematodes) are also biopesticides (e.g. Waage, 1997). While this is undoubtedly correct as a scientific description, most regulatory agencies reject this view. Governments have tended to use different legislative approaches for macro biocontrol agents and microbial agents/ natural products.

Genetic methods
The mass production and release of sterile male insects as a mating disruption strategy has proved very effective against some pests, such as screw-worm and fruit flies. It involves the release of very large numbers of sterilized insects into the vicinity of the crop, so that they will mate with resident insects and so prevent them producing young. The insects to be released are sterilized using irradiation or chemicals. The release ratio ranges from 10:1 up to 100:1 sterile to wild insects. The technique is expensive, mainly because large numbers of insects must be reared and it

usually requires releases over large areas. It has been used successfully to eradicate the screw-worm fly (*Cochliomyia hominivorax*) in areas of North America. There have also been many successes in controlling species of fruit flies, most particularly the medfly (*Ceratitis capitata*) (Hendrichs *et al.*, 2002).

Because this book focuses on biopesticides, it is not our intention to go into detail about other, alternative methods of pest control. However, crop protection practitioners, regulators and others who work with biopesticides need to have a thorough understanding of other crop protection methods in order that the best use can be made of biopesticides within IPM. For further information on crop protection technologies, readers should consult Smith (1995), Agrios (2005) and Radcliffe *et al.* (2008).

It is worth pointing out some potential flaws and pitfalls in the way crop protection agents are currently categorized. Some of the categories listed above clearly overlap. Microbial biopesticides, for example, are also biological control agents. New synthetic pesticides may be based on the molecular structure of natural compounds. Unfortunately, there are no universally agreed definitions for many of the crop protection approaches used around the world. Take biological control, for example. In this book, we are sticking with a 'traditional' view of biological control as requiring the use of living organisms. However, some authors consider it to include genes and gene products. Thus the US National Academy of Sciences (1988) describes biological control as 'the use of natural or modified organisms, genes or gene products to reduce the effects of undesirable organisms (pests) and to favour desirable organisms such as crops, trees, animals and beneficial insects and micro organisms'. According to this definition, biological control includes agents such as GM organisms and natural products that have pesticidal properties (e.g. neem oil) or can modify pest behaviour (e.g. insect pheromones). Crop varieties modified genetically to express the insecticidal protein from *Bacillus thuringiensis* could be categorized as plant breeding, as a biopesticide or as biological control. Some biological control experts are happy with the expanded definition of biological control, while others think it loses the original emphasis on applying theories about interactions between species and their natural enemies. Obtaining universally agreed definitions is important because it affects how governments regulate crop protection agents and how scientists and practitioners develop them.

Integrated Pest Management in Practice

The first aim of IPM must be to use chemical pesticides at a sustainable level without causing unacceptable loss of crop yields. However, in order to keep up with the increase in global food production that will be required in the next 20 years, IPM will need to be capable of moving to the next level fairly rapidly, namely to enable yields to be increased while still keeping use of pesticides and other inputs at sustainable levels. Can IPM deliver? In a groundbreaking study, Pretty (2008) analysed data from 62 IPM projects in 26 industrialized and developing countries, covering over 5 million farm households farming 25 million ha. His analysis indicates that IPM is successful in most cases. Only one out of the 62 IPM projects resulted in an increase in pesticide use and an associated decline in yields. Over 60% of the IPM projects resulted in a reduction in pesticide use (average reduction of 75%) and an increase in yields (average yield increase of 40%). An additional 15% of projects resulted in an increase of both yield (average 45% increase) and pesticide use (average 20% increase). These were mainly conservation farming projects that incorporated zero tillage to conserve soils and reduce water pollution and they tended to result in greater use of herbicides for weed management. Approximately 20% of the IPM projects resulted in a slight reduction in yield (average 5% reduction) and lowered pesticide use (average 60% reduction). These mainly consisted of cereal production projects in Europe.

So if IPM is successful, are many farmers using it on the ground? Figures for the estimated area of crops currently under supervised control and IPM in Europe for different crop types are given in Table 2.6 (Bale *et al.*, 2008).

Table 2.6. Supervised and integrated control programmes used in Europe. (From Bale *et al.*, 2008.)

Crop	Type	Elements	Area under IPM in Europe/reduction in pesticides on that area
Field vegetables	Supervised	Monitoring; sampling; disease-resistant crop varieties	5% of total area/ 20–80% reduction
Cereals	Supervised	Monitoring; sampling; forecasting; resistant crop varieties	10% of total area/ 20–50% reduction
Maize	Integrated	Mechanical weeding; resistant crop varieties; biological control of insects	4% of total area/ 30–50% reduction
Vineyards	Integrated	Biological control of insects and mites; disease-resistant crop varieties; pheromone mating disruption	20% of total area/ 30–50% reduction
Olives	Integrated	Cultural control; biological control of insects; disease- and insect-resistant crop varieties; monitoring; sampling; pheromones	Very limited
Orchards	Supervised	Monitoring; sampling; selective pesticides	15% of total area/ 30% reduction
	Integrated	Monitoring; sampling; pheromone mating disruption; biological control; disease-resistant crop varieties	7% of total area/ 50% reduction
Greenhouse vegetables	Integrated	Monitoring; sampling; biological control of insects, mites and diseases; disease-resistant crop varieties; selective pesticides	30% of total area/ 50–99% reduction

The data show that few farmers use IPM, although there are exceptions, such as greenhouse growers in the UK and the Netherlands. A lot of research has been done on IPM since the mid-1980s, so the fact that most farmers have yet to adopt it raises major issues about how research can be translated into farmer practice. The degree to which IPM has been adopted and is successful (i.e. leads to reductions in pesticide use and still maintains yield) varies significantly according to the type of crop grown. Crop protection tactics have to be chosen that are appropriate not only to the crop plant, but also to the physical area of production, the type and number of pests, and the physical, chemical and biological environments. Different tactics must also be capable of being integrated, and they must be affordable for the farmer. It is not surprising that IPM is most advanced in high-value glasshouse crops, such as fresh salads or ornamental plants. For these crops, growers can afford to spend comparatively large sums on crop protection (including labour-intensive methods such as spot-treating individual plants). There are also strong driving forces to make IPM work in the form of a significant pesticide resistance problem, extremely high quality standards from retailers, and pressure from consumers and others to use fewer pesticides. In contrast, producers of outdoor broad-acre crops such as cereals, potatoes and cotton have much less to spend on pest management per unit area of crop, and have to work in a much more challenging physical and ecological environment.

Integrated pest management in greenhouse crops

The global area of crop production done in greenhouses (which include glasshouses, polythene tunnels and other forms of protective structure) is estimated at around 2.4 million ha, of which around 45,000 ha is done in glasshouses (van Lenteren, 2006). Greenhouse growers are able to produce high-value crops on a small area of land. Unfortunately, greenhouse crops also provide an excellent environment for pest insects, mites and plant pathogens. Pesticide resistance evolved in some key glasshouse pests as long ago as the 1960s, prompting the early development of biocontrol. This was followed by the widespread adoption of bumblebees for pollination, which required growers to stop using broad-spectrum insecticides. Growers of greenhouse edible crops are also under severe pressure from retailers to deliver produce with zero detectable pesticide residues because of consumer concerns about residue safety (Pilkington et al., 2010).

Some very sophisticated and effective IPM programmes have been developed for glasshouse crops (Table 2.7). These were mainly instigated through publicly funded research and involved close working between research scientists, growers and industry. We would class these as Level Four IPM according to the Prokopy scale (Prokopy, 1993). In Europe, IPM is used in over 90% of glasshouse tomato, cucumber and sweet pepper production in the Netherlands (van Lenteren, 2000) and is standard practice for glasshouse crops in the UK. In Almeria, Spain, the area under IPM has increased from just 250 ha in 2005 to around 7000 ha in 2008, while the proportion of the Dutch

Table 2.7. Integrated pest and disease management programme as applied in tomato in Europe. (From van Lenteren, 2000.)

Pest/disease	Method used to prevent or control pest/disease
Pests	
Whiteflies (*Bemisia tabaci, Trialeurodes vaporariorum*)	Parasitoids: *Encarsia, Eretmocerus* Predators: *Macrolophus* Pathogens: *Lecanicillium, Paecilomyces, Aschersonia*
Spider mite (*Tetranychus urticae*)	Predators: *Phytoseiulus*
Leaf miners (*Liriomyza bryoniae, Liriomyza trifolii* and *Liriomyza huidobrensis*)	Parasitoids: *Dacnusa, Diglyphus, Opius*, natural control
Lepidoptera (e.g. *Chrysodeixis chalcites, Lacanobia oleracea, Spodoptera littoralis*)	Parasitoids: *Trichogramma* Pathogens: *Bacillus thuringiensis*
Aphids (e.g. *Myzus persicae, Aphis gossypii, Macrosiphum euphorbiae*)	Parasitoids: *Aphidius, Aphelinus* Predators: *Aphidoletes*, natural control
Nematodes (e.g. *Meloidogyne* spp.)	Resistant and tolerant cultivars, soil-less culture
Diseases	
Grey mould (*Botrytis cinerea*)	Climate management, mechanical control, selective fungicides
Leaf mould (*Fulvia = Cladosporium*)	Resistant cultivars, climate management
Mildew (*Oidium lycopersicon*)	Selective fungicides
Fusarium wilt (*Fusarium oxysporum lycopersici*)	Resistant cultivars, soil-less cultures
Fusarium foot rot (*Fusarium oxysporum radicis/lycopersici*)	Resistant cultivars, soil-less culture, hygiene
Verticillium wilt (*Verticillium dahliae*)	Pathogen-free seed, tolerant cultivars, climate control, soil-less culture
Bacterial canker (*Clavibacter michiganesis*)	Pathogen-free seed, soil-less culture
Several viral diseases	Resistant cultivars, soil-less culture, hygiene, weed management, vector control
Pollination	Bumblebees or bees

chrysanthemum crop grown under IPM has increased from just 1% in 2002 to 80% in 2007 (R. GreatRex, Syngenta Bioline, personal communication). Glasshouse IPM uses a combination of biological and physical controls, selective pesticides and resistant varieties, with very careful monitoring of pest populations to determine when they have passed the economic threshold. This system requires considerable knowledge on behalf of the grower but it has been adopted widely because it has clear benefits. These include reliable pest control, lack of phytotoxic effects, better fruit set and the fact that – because synthetic pesticides are used sparingly – operators do not have to be excluded from the glasshouse after spray applications so often. Unlike for synthetic pesticides, there is no statutory interval between the application of predators/parasitoids and harvesting the crop, while for biopesticides the harvest interval tends to be very short. This means that these agents can be used right up to harvest, which is a considerable benefit when the produce quality standard demanded by retailers is so high. Most of the biological control used in glasshouses is concerned with managing insect and mite pests. Some microbial biopesticides are available against plant pathogens and can be integrated with selective pesticides but a greater range of products is required. Many of the main plant diseases are tackled currently using resistant crop cultivars, although an increasing number of effective microbial antagonists are becoming available. Typically, selective synthetic chemical insecticides and acaricides will be used at the start of the season as a clean-up for insect and mite pests before switching to inundative applications of predators, parasitoids, parasitic nematodes and insect pathogens. Short-persistence pesticides are used on an at-need basis to knock back pest populations if they start to outstrip the ability of the biological control agents to regulate them.

Integrated pest management in field vegetable crops

Retailers and consumers in the industrialized countries increasingly demand vegetables that are relatively uniform in size and shape and free from blemishes and pest-related debris. Field vegetables generally have a higher value than cereals, but a lower value than glasshouse crops. Because of the emphasis on quality, the risk of producing an unmarketable crop is relatively great. Vegetable growers are, therefore, generally risk averse. Pesticides remain the mainstay of crop protection for many vegetable crops. Most of the crop protection is based around supervised spraying of pesticides or there may be some integration of different tactics to control an individual pest species, which we would class as Level One IPM. Other components of IPM that are in use in field vegetables include crop rotation (with some exceptions), good crop management (removal of plant residues, application of fertilizers and irrigation) and adjustment of planting or harvesting dates (USDA, 2001). This may involve the use of pest or disease forecasts (Gilles *et al.*, 2004). Growers walk their crops regularly and make decisions based on their findings. In the USA, research-based economic thresholds are available for a range of crops; however, they are used less often in Europe (Collier and Finch, 2007). Pest- and disease-resistant cultivars are grown where available and if the cultivars meet other market requirements. For example, lettuce with resistance to downy mildew is grown widely. Use of physical methods of pest control is increasing. Mechanical weed control can be effective and covers or mulches have been used to control pest insects and weeds; however, most growers still rely on chemical herbicides for weed control, especially glyphosate, which has a very good human safety and environmental profile. Biological control with microbials or arthropods is relatively undeveloped. This is due to a combination of factors: lack of environmental control and problems of confining released natural enemies to the crop, the relatively high cost of biological control methods compared with the value of the crop, reduced and variable efficacy compared with pesticides, and the limited research and development input in this area. However, there are some exceptional examples of Level Two IPM that indicate how things could be done across much of the vegetable crops sector in the future. Notably, these IPM systems have been driven by very much the same set of conditions that has forced the development of IPM in glasshouse crops. IPM programmes for the production of tomato crops in Florida, for example, were developed in response to pesticide resistance and severe secondary pest problems with *Liriomyza* leaf miners caused by the use of broad-spectrum insecticides sprayed against *Spodoptera* spp. caterpillars. The IPM system that was developed uses pest monitoring and economic thresholds to inform the application of foliar sprays of the bacterial insecticide *Bacillus thuringiensis* (Bt), which is highly selective and preserves natural enemies (Walker et al., 2003).

Integrated pest management in orchard crops

Retailers in the industrialized countries also have high quality standards for orchard crops. Up until the 1970s, pest management in orchard crops relied on calendar sprays of broad-spectrum pesticides, but pesticide resistance became common in many of the key pests. The IPM strategies currently used in orchards are a step forward from supervised pesticide spray programmes developed in the 1970s and 1980s, and are largely based on stopping sprays of broad-spectrum pesticides and allowing natural enemies to re-establish, such as the predatory phytoseiid mite, *Typhlodromus pyri*, which is a key predator of the two-spotted spider mite, *T. urticae*, in apple orchards (Blommers, 1994). The level of sophistication of the pest management programme can vary widely from grower to grower, but in the last decade there has been an increasing awareness of the benefits of adopting IPM by growers in many of the main producer countries. Pesticide sprays are still essential tools for the majority of farmers, especially for the control of diseases such as scab and powdery mildew, but progressive growers are using disease forecasts in order to target spray applications better (MacHardy, 2000). The development of some sophisticated decision support tools (pest thresholds, monitoring and models) has allowed the targeted application of selective insecticides, insect pheromones for mating disruption, microbial control agents and predators and parasitoids against aphids and moth pests without harming predatory mites (Solomon, 1987; MacHardy, 2000; Cross and Berrie, 2006). This is combined with careful management of sites away from the tree where pests overwinter or which act as pest reservoirs, such as the soil, border areas and vegetation beneath the tree canopy. Although monitoring has costs associated with it, successful growers find that savings on reduced application of pesticides and other control agents offset these. There is still much work to be done. For example, effective biopesticides are required for aphid pests and plant pathogens, many of the main commercial crop varieties have low levels of disease resistance, and there is a requirement for better integration of the different IPM systems used for individual pests and diseases.

Integrated pest management in broad-acre crops

There have been significant problems since the 1950s with the over-use of synthetic chemical pesticides in broad-acre crops including cereals, maize and cotton. In some well-documented cases, such as cotton production in the Canete valley of Peru, this has included complete failure to control insect pests following the rapid evolution of pesticide resistance (Thacker, 2002). Broad-acre crops generally have high economic thresholds for pest management interventions, meaning that low densities of many pests can be tolerated (the exceptions are for pre- and postharvest fungal diseases, some of which pose a serious hazard to human health). There is, therefore, good scope for IPM, but it is not yet being used widely. For example, it is estimated that only 10% of cereal production is done under IPM (van Lenteren, 2008). The majority of schemes are based on host-plant resistance and forecasting/monitoring to guide pesticide applications, rather than on biological control. An exception is the use of the parasitoid *Trichogramma* to control outbreaks of the European corn borer, *Ostrinia nubilalis*, on maize (Burn, 1987). As we will discuss later in this book, since the beginning of the 21st century maize and cotton production has been revolutionized by the commercialization of GM varieties that express a gene from the entomopathogenic bacterium *B. thuringiensis*, which codes for a selective, insecticidal protein that kills lepidopteran pests. This protein is safe for mammals and other vertebrates and does not affect the arthropod natural enemies that occur in maize and cotton fields. It has been taken up widely by farmers in many regions of the world with the exception of Europe, where there are deep-seated societal concerns about GM crops. Similarly, weed control has changed greatly since the introduction of GM, herbicide-resistant crops in 1996. Most of these crops have been engineered to be resistant to glyphosate. Over 80 million ha of herbicide-resistant crops were grown in 2006 (Gianessi, 2008). Most of these crops are maize, canola (oilseed rape), soybean or cotton. For example, 80% of US-grown cotton is now resistant to glyphosate. Because glyphosate is a broad-spectrum herbicide, the development of glyphosate-resistant crops has simplified weed control by enabling the farmer to use just the one type of herbicide during production and lowering the number of sprays that have to be applied including the use of pre-emergence herbicides. Glyphosate has a low toxicity to mammals and other vertebrates and, because it is readily adsorbed by soil, there is little risk of it leaching into groundwater (Bayliss, 2000). Even though glyphosate-resistant crop seed is sold to farmers at a premium price, this is offset by reduced costs from herbicides and mechanical weed control and higher yields due to better weed management. The global cumulative farm income benefit from herbicide-resistant crops for the period 1996–2005 is estimated at over US$19 billion (Gianessi, 2008). Resistance to glyphosate is still relatively rare considering the extent to which it is used, and so far only 18 resistant weed species have been documented (WeedScience.com, 2010). However, given the large-scale uptake of glyphosate-resistant crops, there are concerns that there will be increased selection pressure for resistance in weed populations. Hence the recent development of glyphosate resistance in two important weeds of soybean, *Sorghum halepense* and *Euphorbia heterophylla*, in glyphosate-resistant soybean fields in South America is something of a worry (Vila-Aiub *et al.*, 2008).

In Europe, the use of conservation biological control methods, such as wildflower headlands and beetle banks, is promoted through various agri-environment schemes but in the UK, for example, most of these have had a very low uptake (Boatman *et al.*, 2007). There is greater use of IPM techniques within organic arable production, but even here there is limited use of biological control, with the majority of integrated crop management schemes relying on improved monitoring and reduced applications of approved chemical controls (MacKerron *et al.*, 1999). This is not to say that IPM cannot be made to work on broad-acre crops. If the correct research is financed, done, and translated into practice, then IPM is possible. As an example of what can be achieved, a highly successful programme for management of the whitefly *B. tabaci* was implemented on cotton in Arizona in 1995 (Naranjo and Ellsworth, 2009a,b). The programme relies on decision support tools based on pest sampling and economic thresholds for the targeted application of selective insecticides, including insect growth regulators such as buprofezin and pyriproxyfen. The use of these insecticides controls outbreak populations of the whitefly and preserves natural populations of arthropod predators, which typically consist of about 20 different taxa and which suppress the whitefly population below the economic threshold for up to a year after spraying (Naranjo and Ellsworth, 2009a,b). The system has now been integrated with the planting of GM Bt cotton for control of the principal lepidopteran pest, pink bollworm, *Pectinophora gossypiella*. It has resulted in a 70% reduction in the use of foliar insecticides together with control cost savings of over US$200 million in the 14 years from 1995 to 2009.

Integrated pest management in developing countries

Farmers in developing nations face different constraints and opportunities compared with their counterparts farming in industrialized countries. Subsistence farmers in sub-Saharan Africa, Asia and Latin America, for example, operate on narrow economic margins in an ecological environment where crop production is severely constrained by environmental factors such as water availability and soil fertility. These farmers often cannot afford pesticides and other inputs and it is important that improvements are made to basic agronomy before IPM can be considered seriously (Way and van Emden, 2000). This is not to say that IPM has no role here, but it must be suitable and workable for local needs. Where farmers in developing nations are less constrained by access to resources, then the full spectrum of IPM tactics outlined earlier in this book has much to offer providing that they are affordable. Self-evidently, IPM must be tailored to the needs of local farming communities; it is not a case of simply transferring an IPM model from the Global North to the Global South. That, very clearly, could never be justified on economic, environmental, agronomic or social grounds. Although it is not the focus of this book, we must not lose sight of the fact that there is a deep connection between the development of IPM and issues of social justice, fair trade, the development of an effective policy environment (many developing countries lack clear policies on IPM), access to loans and subsidies for IPM and other tools, and improvements to the market infrastructure such as better roads and transport for agricultural produce.

Historically, many farmers in developing countries have suffered the same problems associated with uncontrolled pesticide use as elsewhere, namely the rapid evolution of pesticide resistance, resurgence, problems of secondary pests and environmental pollution. The intensive use of broad-spectrum pesticides for rice production in South-east Asia, for example, which was associated with the adoption of new high-yielding varieties in the Green Revolution of the 1960s–1970s, created major problems of resurgence and secondary pest outbreaks. In general, there is still an over-reliance on 'spray-and-pray' approaches in many developing countries. The agricultural science research base in these countries is generally highly capable and adept but suffers from a lack of investment. Nevertheless, there have been important and exciting successes in IPM in developing nations since the 1970s (Way and van Emden, 2000). In South-east Asia, the development of resistant rice varieties, combined with publicly funded pest surveillance and IPM training through farmer field schools and participatory experiments, has allowed significant reductions to be made in pesticide applications (Thacker, 2002). In India, an IPM programme based on insecticide resistance management in cotton has operated over ten different states since 1992. The programme is based on targeted applications of selective insecticides to preserve natural enemies and pesticide rotation, done according to the same ecological principles used in the Arizona cotton IPM programme discussed previously. The Indian system has resulted in a 30% reduction in the amount of pesticide applied to the crop, and although it has not increased yields, it has increased the average net financial returns of farmers participating in the programme by 11% (Peshin et al., 2009). In Africa, classical biological control has proved effective against some invasive pests, most notably on cassava and maize, while intercropping and the development of the 'push–pull' strategy are proving successful for maize and sorghum production. Developing countries are also now investing in home-grown biotechnology research to produce new crop varieties with improved resistance to arthropods and plant pathogens. Low labour costs have also enabled the mass production of arthropod natural enemies and microbial biopesticides to be done economically using relatively unsophisticated but robust methods. Indeed, the industrialized countries have much to learn from the developing nations about the effective implementation of biopesticides and biological control. Thus, while much of the English-language published research on entomopathogenic fungi comes from Europe and North America, the practical use of fungal bioinsecticides and bioacaricides has really been pioneered in Central and South America; 50% of all such products have been developed and commercialized here (Faria and Wraight, 2007). A critical point to note is that while IPM initiatives are often done according to a top-down approach, in which pest problems are identified and researched by scientists and the results are then transferred to farmers, some of the most effective IPM initiatives in developing nations have been more bottom-up, with farmers and researchers working together to identify the main pest issues. Training and education have then been provided for farmers, which has enabled them to develop their own solutions catered to their particular needs and making best use of traditional agricultural systems (for an excellent discussion of this topic in relation to biopesticides, the reader is referred to Rodriguez and Niemeyer, 2005). A highly important development for Africa and Latin America has been the creation of new export markets to Europe and North America, particularly for smallholders growing under contract to multinational retailers. In Kenya, for example, 16% of the country's smallholders now grow horticultural crops under contract for export (Nwilene et al., 2008). The requirement for these growers to comply with international standards for food safety and environmental protection (e.g. the EurepGAP standards of the EU) has led to significantly greater focus on IPM done to European and North American standards.

Using Alternatives to Synthetic Chemical Pesticides in Integrated Pest Management

As we look to the future, it is clear that synthetic chemical pesticides will remain a critical tactic in crop protection, but they should be used only when necessary in order to reduce the chances of resistance developing and to minimize negative effects on non-target organisms and the wider environment. This will inevitably mean that alternative crop protection methods will have to be used more. This includes biological controls, biopesticides, cultural methods and a wider use of resistant varieties. It will inevitably require greater consideration to be given to tactics that some people are uncomfortable with for ethical or moral reasons, such as GM crops. We argue that the debate about the future of IPM must be framed in terms of the best practices that can be adopted from all farming systems to make crop protection more sustainable. Thus, according to Pretty (2008), sustainable agriculture 'does not mean ruling out any technologies or practices on ideological grounds. If a technology works to improve productivity for farmers and does not cause undue harm to the environment, then it is likely to have some sustainability benefits.'

There are a number of caveats about alternative, non-pesticide methods of pest control that must be borne in mind: (i) these methods are generally not like-for-like replacements for synthetic chemical pesticides; (ii) they have different modes of action and, thus, different strengths and weaknesses; (iii) many have lower efficacy than synthetic pesticides, are slower to act or are vulnerable to breakdown; (iv) many are affected by environmental conditions; (v) they often cost more to buy than synthetic pesticides and they require specialist knowledge to use them; (vi) there are also many remaining scientific and technical challenges to their development, not helped by under-investment in crop protection research; and (vii) some of the features that make them attractive from an environmental standpoint can be a problem in economic terms. For example, the narrow prey spectrum of many biological control agents means that they are appropriate only for niche markets, with individual agents often earning less than €1 million per product per annum, which discourages companies from developing them. To quote Gelernter (2005): 'The features that made most BCPs [biological control products] so attractive from the standpoint of environmental and human safety also acted to limit the number of markets in which they were effective.'

Many of the perceived 'weaknesses' of non-pesticide agents are based on comparison with the performance of synthetic chemicals. The unsuitable adoption of a chemical pesticide development model for these agents can lead to false and unrealistic expectations of chemical-like performance. However, the very fact that 'alternative' crop protection methods do not function in the same way as synthetic chemical pesticides means that they can be used to compensate for the weaknesses of the latter. Biological control agents are able to reproduce within the pest population, giving various levels of self-perpetuating control. Many biological control agents are also able to actively locate their prey, making them highly useful where pests occupy cryptic habitats. Biological control, plant breeding and cultural controls generally do not require the same level of protective equipment to be used as for chemical pesticides. The lack of a toxic residue on crops and, therefore, the absence of a preharvest application interval, can be a real benefit. Using knowledge of the causes of pest outbreaks it is possible to use conservation management, plant breeding, physical methods and cultural techniques to help prevent the build-up of pest populations in a 'total systems' approach (Lewis et al., 1997). Moreover, the wide variety of crop protection methods available means that different combinations can be put together in complementary ways. This is vital if we are to achieve effective, sustainable pest management. There is ample evidence that the various non-chemical pesticide methods can make valuable contributions to crop protection as part of IPM. In some situations, a combination of these methods may be able to replace synthetic pesticides, for example where a pest has developed pesticide resistance. But in most cases the most practical way forward is to use them in combination with chemicals in a fully integrated programme.

Table 2.8. Estimated share of 2004 global product sales for biologically based control agents. (From Gelernter, 2005.)

Class of product	Market share (%)
Microbial biopesticides	65–70
Beneficial macroorganisms	15–16
Semiochemicals	10–19
Botanicals	4–8
Microbial soil and plant enhancers	1–2

Compared with synthetic chemical pesticides, the current market size for 'alternative' crop protection products is small. For example, the global market for biologically based plant protection products (by which we mean biopesticides plus predators and parasitoids) represents just 2.5% of the global chemical pesticide market, and is currently valued at €60 million (Evans, 2008). At present biopesticide products based on microorganisms make up the majority of the market share for biologically based control agents (Table 2.8) (Gelernter, 2005). However, market size alone undoubtedly underestimates the contribution these methods can make to improve sustainability, for example by reducing the resistance pressure on chemical pesticides. And as we shall read in Chapter 3, the market for biopesticides and other alternatives is increasing. With this in mind, we shall now turn our attention to biopesticides: what are they, what are their properties, and how can they be used in crop protection?

References

Agrios, G.N. (2005) *Plant Pathology*, 5th edn. Elsevier Academic Press, Burlington, Massachusetts.

Alyokhin, A., Dively, G., Patterson, M., Castaldo, C., Rogers, D., Mahoney, M. and Wollam, J. (2007) Resistance and cross-resistance to imidacloprid and thiamethoxam in the Colorado potato beetle *Leptinotarsa decemlineata*. *Pest Management Science* 563, 32–41.

Bajwa, W.I. and Kogan, M. (2002) *Compendium of IPM Definitions (CID) – What is IPM and How is it Defined in the Worldwide Literature?* IPPC Publication No. 998. Integrated Plant Protection Center, Oregon State University, Corvallis, Oregon.

Bale, J.S., van Lenteren, J.C. and Bigler, F. (2008) Biological control and sustainable food production. *Philosophical Transactions of the Royal Society B* 363, 761–776.

Bayliss, A.D. (2000) Why glyphosate is a global herbicide: strengths, weaknesses and prospects. *Pest Management Science* 56, 299–308.

Begon, M., Townsend, C.R. and Harper, J.L. (2006) *Ecology: From Individuals to Ecosystems*. Blackwell Publishing Ltd, Malden, Massachusetts.

Bellows, T.S. and Fisher, T.W. (eds) (1999) *Handbook of Biological Control*. Academic Press, San Diego, California.

Blommers, L.H.M. (1994) Integrated pest management in European apple orchards. *Annual Review of Entomology* 39, 213–241.

Boatman, N.D., Jones, N.E., Garthwaite, D. and Pietravalle, S. (2007) Option uptake in entry level scheme agreements in England. *Aspects of Applied Biology* 81, 309–316.

Bond, W. and Grundy, A.C. (2001) Non-chemical weed management in organic farming systems. *Weed Research* 41, 383–405.

Bondada, B.R., Oosterhuis, D.M., Tugwell, N.P. and Kim, K.S. (1995) Physiological and cytological studies of two spotted spider mite, *Tetranychus urticae* K., injury in cotton. *Southwestern Entomologist* 20, 171–180.

Bonman, J.M., Khush, G.S. and Nelson, R.J. (1992) Breeding rice for resistance to pests. *Annual Review of Phytopathology* 30, 507–528.

Burn, A.J. (1987) Cereal crops. In: Burn, A.J., Coaker, T.H. and Jepson, P.C. (eds) *Integrated Pest Management*. Academic Press, London, pp. 209–256.

Carlile, M.J., Watkinson, S.C. and Gooday, G.W. (2001) *The Fungi*. Academic Press, London.

Collard, B.C.Y. and Mackill, D.J. (2008) Marker-assisted selection: an approach for precision plant breeding in the twenty first century. *Philosophical Transactions of the Royal Society B* 363, 557–590.

Collier, R.H. and Finch, S. (2007) IPM case studies: brassicas. In: van Emden, H.F. and Harrington, R. (eds) *Aphids as Crop Pests*. CAB International, Wallingford, UK, pp. 549–572.

Coop, L.B. and Croft, B.A. (1995) *Neoseiulus fallacis* dispersal and biological control of *Tetranychus urticae* following minimal inoculations into a strawberry field. *Experimental and Applied Acarology* 19, 31–43.

Copping, L.G. and Menn, J.J. (2000) Biopesticides: a review of their action, applications and efficacy. *Pest Management Science* 56, 651–676.

Cranham, J.E. and Helle, W. (1985) Pesticide resistance in Tetranychidae. In: Helle, W. and Sabelis, M.W. (eds) *Spider Mites. Their Biology, Natural Enemies and Control*, Vol. 1B. Elsevier, Amsterdam, pp. 405–421.

Croft, B.A. and Slone, D.H. (1998) Perturbation of regulated apple mites: immigration and pesticide effects on outbreaks of *Panonychus ulmi* and associated mites (Acari: Tetranychidae, Eriophyidae, Phytoseiidae and Stigaeidae). *Environmental Entomology* 27, 1548–1556.

Cross, J. and Berrie, A. (2006) The challenges of developing IPM programmes for soft fruit crops that eliminate reportable pesticide residues. *Journal of Fruit and Ornamental Plant Research* 14(Suppl. 3), 49–59.

DAISIE (2009) Delivering Alien Invasive Species Inventories for Europe. European Commission Sixth Framework Programme (Contract Number: SSPI-CT-2003-511202). Available at: http://www.europe-aliens.org/ (accessed 14 October 2009).

Dent, D. (2000) *Insect Pest Management*. CAB International, Wallingford, UK.

Desneux, N., Decourtye, A. and Delpuech, J.-M. (2007) The sublethal effects of pesticides on beneficial arthropods. *Annual Review of Entomology* 52, 81–106.

de Waard, M.A., Georgopoulos, S.G., Hollomon, D.W., Ishii, H., Leroux, P., Ragsdale, N.N. and Schwinn, F.J. (1993) Chemical control of plant diseases: problems and prospects. *Annual Review of Phytopathology* 31, 403–421.

Duke, S.O. and Powles, S.B. (2008) Glyphosate: a once-in-a-century herbicide. *Pest Management Science* 64, 319–325.

Dwyer, G., Dushoff, J. and Yee, S.H. (2004) The combined effects of pathogens and predators on insect outbreaks. *Nature* 430, 341–345.

Eilenberg, J., Hajek, A. and Lomer, C. (2001) Suggestions for unifying the terminology in biological control. *BioControl* 46, 387–400.

Ellis, P.R., Hardman, J.A., Cole, R.A. and Phelps, K. (1987) The complementary effects of plant resistance and the choice of sowing and harvest times in reducing carrot fly (*Psila rosae*) damage to carrots. *Annals of Applied Biology* 111, 415–424.

Evans, J. (2008) Biopesticides: from cult to mainstream. *Agrow* October 2008, 11–14. Available at: http://www.agrow.com

FAO (2010) Staple foods: What do people eat? Available at: http://www.fao.org/docrep/u8480e/u8480e07.htm (accessed April 2010).

Faria, M.R. and Wraight, S.P. (2007) Mycoinsecticides and mycoacaricides: a comprehensive list with worldwide coverage and international classification of formulation types. *Biological Control* 43, 237–256.

Feeny, P. (1976) Plant apparency and chemical defense. *Recent Advances in Phytochemistry* 10, 1–40.

Finch, S. and Collier, R.H. (2000) Host-plant selection by insects – a theory based on 'appropriate/inappropriate landings' by pest insects of cruciferous plants. *Entomologia Experimentalis et Applicata* 96, 91–102.

Flint, M.L. and van den Bosch, R. (1981) *Introduction to Integrated Pest Management.* Plenum Press, New York.

Fry, W.E., Goodwin, S.B., Matuszak, J.M., Spielman, L.J. and Milgroom, M.G. (1992) Population genetics and intercontinental migrations of *Phytophthora infestans. Annual Review of Phytopathology* 30, 107–129.

Gelernter, W.D. (2005) Biological control products in a changing landscape. In: *The BCPC International Congress Proceedings: 2005*, Vol. 1. British Crop Protection Council, Alton, UK, pp. 293–300.

Gianessi, L.P. (2008) Economic impacts of glyphosate-resistant crops. *Pest Management Science* 64, 346–352.

Gilles, T., Phelps, K., Clarkson, J.P. and Kennedy, R. (2004) Development of MILION-CAST, an improved model for predicting downy mildew sporulation on onions. *Plant Disease* 88, 695–702.

Godan, D. (1983) *Pest Slugs and Snails. Biology and Control.* Springer-Verlag, Berlin.

Goodwin, S.B., Cohen, B.A. and Fry, W.E. (1994) Panglobal distribution of a single clonal lineage of the Irish potato famine fungus. *Proceedings of the National Academy of Sciences USA* 91, 11591–11595.

Grafton-Cardwell, E.E., Lee, J.E., Stewart, J.R. and Olsen, K.D. (2006) Role of two insect growth regulators in Integrated Pest Management of citrus scales. *Journal of Economic Entomology* 99, 733–744.

Gray, M.E., Sappington, T.W., Miller, N.J., Moeser, J. and Bohn, M.O. (2009) Adaptation and invasiveness of western corn rootworm: intensifying research on a worsening pest. *Annual Review of Entomology* 54, 303–321.

Green, M.B., Hartley, G.S. and West, T.F. (1987) *Chemicals for Crop Improvement and Pest Management.* Pergamon Press, Oxford, UK.

Grunwald, N.J. and Flier, W.G. (2005) The biology of *Phytophthora infestans* at its center of origin. *Annual Review of Phytopathology* 43, 171–190.

Gullan, P.J. and Cranston, P.S. (1994) *The Insects: an Outline of Entomology.* Chapman and Hall, London.

Hajek, A. (2004) *Natural Enemies: an Introduction to Biological Control.* Cambridge University Press, Cambridge.

Hardman, J.M., Smith, R.F. and Bent, E. (1995) Effects of different integrated pest management programs on biological control of mites on apple by predatory mites (Acari) in Nova Scotia. *Environmental Entomology* 24, 125–142.

Hare, J.D. (1990) Ecology and management of the Colorado potato beetle. *Annual Review of Entomology* 35, 81–100.

Helle, W. and Sabelis, M.W. (eds) (1985a) *Spider Mites. Their Biology, Natural Enemies and Control*, Vol. 1A. Elsevier, Amsterdam.

Helle, W. and Sabelis, M.W. (eds) (1985b) *Spider Mites. Their Biology, Natural Enemies and Control*, Vol. 1B. Elsevier, Amsterdam.

Hendrichs, J., Robinson, A.S., Cayol, J.P. and Enkerlin, W. (2002) Medfly areawide sterile insect technique programmes for prevention, suppression or eradication: the importance of mating behavior studies. *Florida Entomologist* 85, 1–13.

Hesketh, H., Roy, H.E., Eilenberg, J., Pell, J.K. and Hails, R.S. (2009) Challenges in modelling complexity of fungal entomopathogens in semi-natural populations of insects. *BioControl* 55, 55–73.

Hill, D.S. (1994) *Agricultural Entomology.* Timber Press, Portland, Oregon.

Hill, D.S. (1997) *The Economic Importance of Insects.* Chapman and Hall, London.

Hussey, N.W. and Scopes, N.E.A. (1985) Control of Tetranychidae in crops: greenhouse vegetables (Britain). In: Helle, W. and Sabelis, M.W. (eds) *Spider Mites. Their Biology, Natural Enemies and Control*, Vol. 1B. Elsevier, Amsterdam, pp. 285–297.

Knott, C.M. (2002) Weed control in other arable and field vegetable crops. In: Naylor, R.E.L. (ed.) *Weed Management Handbook*, 9th edn. Blackwell Science Ltd, Oxford, UK, pp. 359–398.

Kogan, M. (1998) Integrated pest management: historical perspectives and contemporary developments. *Annual Review of Entomology* 43, 243–270.

Lamb, R.J. (1989) Entomology of oilseed brassica crops. *Annual Review of Entomology* 34, 211–229.

Leino, M. (2007) *Fungal Diseases on Oilseed Rape and Turnip Rape*. Swedish Board of Agriculture, Jönköping, Sweden.

Lester, P.J., Thistlewood, H.M.A. and Harmsen, R. (1998) The effects of refuge size and number on acarine predator–prey dynamics in a pesticide-disturbed apple orchard. *Journal of Applied Ecology* 35, 323–331.

Lewis, W.J, van Lenteren, J.C., Pathak, S.C. and Tumlinson, J.H. (1997) A total systems approach to sustainable pest management. *Proceedings of the National Academy of Sciences USA* 94, 12243–12248.

Lindquist, E.E. (1984) Current theories on the evolution of major groups of Acari and on their relationship with other groups of Arachnida, with consequent implications for their classification. In: Griffiths, D.A. and Bowman, C.E. (eds) *Acarology VI*, Vol. I. John Wiley and Sons, New York, pp. 28–62.

Lindquist, E.E., Sabelis, M.W. and Bruin, J. (eds) (1996) *Eriophyoid Mites, Their Biology, Natural Enemies and Control*. Elsevier, Amsterdam.

Lucas, J.A. (1998) *Plant Pathology and Plant Pathogens*, 3rd edn. Blackwell Science Ltd, Oxford, UK.

MacArthur, R.H. and Wilson, E.O. (1967) *The Theory of Island Biogeography*. Princeton University Press, Princeton, New Jersey.

MacHardy, W.E. (2000) Current status of IPM in apple orchards. *Crop Protection* 19, 801–806.

MacKenzie, D. (2007) Billions at risk from wheat super blight. *New Scientist* 2598, 6–7.

MacKerron, D.K.L., Duncan, J.M., Hillman, J.R., Mackay, G.R., Robinson, D.J., Trudgill, D.L. and Wheatley, R.J. (1999) Organic farming: science and belief. In: Macfarlane-Smith, W.H. and Heilbronn, T.D. (eds) *Scottish Crop Research Institute Annual Report 1998/99*. Scottish Crop Research Institute, Dundee, UK, pp. 60–72. Available at: http://www.scri.ac.uk/scri/file/fullannualreports/annual_report_1999.pdf (accessed 13 July 2010).

McMurtry, J.A. (1985) Control of Tetranychidae in crops: citrus. In: Helle, W. and Sabelis, M.W. (eds) *Spider Mites. Their Biology, Natural Enemies and Control*, Vol. 1B. Elsevier, Amsterdam, pp. 339–347.

Milgroom, M.G., Levin, S.A. and Fry, W.E. (1989) Population genetics theory and fungicide resistance. In: Leonard, K.J. and Fry, W.E. (eds) *Plant Disease Epidemiology*. Vol. 2. *Genetics, Resistance and Management*. McGraw Hill, New York, pp. 340–367.

Millennium Ecosystem Assessment (2005) *Ecosystems and Well-being*. Island Press, Washington, DC.

Minnis, S.T., Haydock, P.P.J., Ibrahim, S.K., Grove, I.G., Evans, K. and Russell, M.D. (2002) Potato cyst nematodes in England and Wales – occurrence and distribution. *Annals of Applied Biology* 40, 187–195.

Mitchell, C.E. and Power, A.G. (2003) Release of invasive plants from fungal and viral pathogens. *Nature* 421, 625–627.

Morton, V. and Staub, T. (2008) *A Short History of Fungicides*. Available at: http://www.apsnet.org/online/feature/fungi/ (accessed 16 April 2010).

Mumford, J.D. and Norton, G.A. (1984) Economics of decision making in pest management. *Annual Review of Entomology* 29, 157–174.

Naranjo, S.E. and Ellsworth, P.C. (2009a) Fifty years of the integrated control concept: moving the model and implementation forward in Arizona. *Pest Management Science* 65, 1267–1286.

Naranjo, S.E. and Ellsworth, P.C. (2009b) The contribution of conservation biological control to integrated control of *Bemisia tabaci* in cotton. *Biological Control* 51, 458–470.

National Academy of Sciences (1988) *Research Briefings 1987: Report of the Research Briefing Panel on Biological Control in Managed Ecosystems*. National Academy Press, Washington, DC.

Naylor, R.E.L. (ed.) (2002) *Weed Management Handbook*. Blackwell Publishing, Oxford, UK.

Naylor, R.E.L. and Drummond, C. (2002) Integrated weed management. In: Naylor, R.E.L. (ed.) *Weed Management Handbook*. Blackwell Publishing, Oxford, UK, pp. 302–310.

Nwilene, F.E., Nwanze, K.F. and Youdeowei, A. (2008) Impact of integrated pest management on food and horticultural crops in Africa. *Entomologia Experimentalis et Applicata* 128, 355–363.

Oerke, E.C. and Dehne, H.W. (2004) Safeguarding production – losses in major crops and the role of crop protection. *Crop Protection* 23, 275–286.

Oerke, E.C., Dehne, H.W., Schoenbeck, F. and Weber, A. (1994) *Crop Production and Crop Protection: Estimated Losses in Major Food and Cash Crops*. Elsevier Science Publishers BV, Amsterdam.

Orr, A. (2003) Integrated pest management for resource-poor African farmers: is the emperor naked? *World Development* 31, 831–845.

Palumbi, S.R. (2001) Humans as the world's greatest evolutionary force. *Science* 293, 1786–1790.

Paoletti, M.G. and Pimentel, D. (2000) Environmental risk of pesticides versus genetic engineering for agricultural pest control. *Journal of Agricultural and Environmental Ethics* 12, 279–303.

Peshin, R., Dhawan, A.K., Kranthi, K.R. and Singh, K. (2009) Evaluation of the benefits of an insecticide resistance management programme in Punjab in India. *International Journal of Pest Management* 55, 207–220.

Pilkington, L.J., Messelink, G., van Lenteren, J.C. and Le Mottee, K. (2010) Protected Biological Control – biological pest management in the greenhouse industry. *Biological Control* 52, 216–220.

Pimentel, D. (1997) *Techniques for Reducing Pesticides: Environmental and Economic Benefits*. John Wiley and Sons, Chichester, UK.

Pimentel, D., McNair, S., Janecka, J., Wightman, J., Simmonds, C., O'Connell, C., Wong, E., Russell, L., Zern, J., Aquino, T. and Tsomondo, T. (2001) Economic and environmental threats of alien plant, animal, and microbe invasions. *Agriculture, Ecosystems & Environment* 84, 1–20.

Pink, D., Bailey, L., McClement, S., Hand, P., Mathas, E., Buchanan-Wollaston, V., Astley, D., King, G. and Teakle, G. (2008) Double haploids, markers and QTL analysis in vegetable brassicas. *Euphytica* 164, 509–514.

Pretty, J. (2008) Agricultural sustainability: concepts, principles and evidence. *Philosophical Transactions of the Royal Society B* 363, 447–465.

Prokopy, R.J. (1993) Stepwise progress toward IPM and sustainable agriculture. *IPM Practitioner* 15, 1–4.

PSD (2008a) Assessment of the impact on crop protection in the UK of the 'cut-off criteria' and substitution provisions in the proposed Regulation of the European Parliament and of the Council concerning the placing of plant protection products in the market. Available at: http://www.pesticides.gov.uk/environment.asp?id=1980andli nk=%2Fuploadedfiles%2FWeb%5FAssets%2FPSD%2FImpact%5Freport%5Ffinal %5F%28May%5F2008%29%2Epdf (accessed 22 November 2008).

PSD (2008b) Plant protection products regulation: agronomic implications of proposals in the EU. Available at: http://www.pesticides.gov.uk/environment.asp?id=1980 andlink=%2Fuploadedfiles%2FWeb%5FAssets%2FPSD%2FPPPR%5FAgronom ic%5Fimplications%5Fof%5Fproposals%5Fin%5Fthe%5FEU%2Epdf (accessed 22 November 2008).

Radcliffe, E.B., Hutchison, W.D. and Cancelado, R.E. (eds) (2008) *Integrated Pest Management: Concepts, Tactics, Strategies and Case Studies*. Cambridge University Press, Cambridge.

Radosevich, S.R., Holt, J.S. and Ghersa, C.M. (2007) *Ecology of Weeds and Invasive Plants: Relationship to Agriculture and Natural Resource Management*, 3rd edn. John Wiley and Sons, Hoboken, New Jersey.

Rhainds, M. and English-Loeb, G. (2003) Testing the resource concentration hypothesis with tarnished plant bug on strawberry: density of hosts and patch size influence the interaction between abundance of nymphs and incidence of damage. *Ecological Entomology* 28, 348–358.

Rodriguez, L.C. and Niemeyer, H.M. (2005) Integrated pest management, semiochemicals and microbial pest-control agents in Latin America agriculture. *Crop Protection* 24, 615–623.

Root, R.B. (1973) Organization of a plant–arthropod association in simple and diverse habitats: the fauna of collards (*Brassica oleracea*). *Ecological Monographs* 43, 95–124.

Skylakakis, G. (1987) Changes in the composition of pathogen populations caused by resistance to fungicides. In: Wolfe, M.S. and Caten, C.E. (eds) *Populations of Plant Pathogens: Their Dynamics and Genetics*. Blackwell Publishing, Oxford, pp. 227–237.

Smith, A.E. (1995) *Handbook of Weed Management Systems*. CRC Press, Boca Raton, Florida.

Smith, K., Evans, D.A. and El-Hiti, G.A. (2008) Role of modern chemistry in sustainable arable crop protection. *Philosophical Transactions of the Royal Society B* 363, 623–637.

Solomon, M.G. (1987) Fruit and hops. In: Burn, A.J., Coaker, T.H. and Jepson, P.C. (eds) *Integrated Pest Management*. Academic Press, London, pp. 329–360.

Solomon-Blackburn, R.M., Stewart, H.E. and Bradshaw, J.E. (2007) Distinguishing major-gene from field resistance to late blight (*Phytophthora infestans*) of potato (*Solanum tuberosum*) and selecting for high levels of field resistance. *Theoretical and Applied Genetics* 115, 141–149.

Sonenshine, D.E. (1993) *Biology of Ticks*, Vol. 2. Oxford University Press, Oxford.

Speranza, C.I., Kiteme, B. and Wiesmann, U. (2008) Droughts and famines: the underlying factors and the causal links among agro-pastoral households in semi-arid Makueni district, Kenya. *Global Environmental Change* 18, 220–233.

Statzner, B., Hildrew, A.G. and Resh, V.H. (2001) Species traits and environmental constraints: entomological research and the history of ecological theory. *Annual Review of Entomology* 46, 291–316.

Strange, R.N. (2003) *Introduction to Plant Pathology*. John Wiley and Sons Ltd, Chichester UK.

Styer, W.E. and Nault, L.R. (1996) Damage and control in eriophyoid mites in crops: corn and grain plants. In: Lindquist, E.E., Sabelis, M.W. and Bruin, J. (eds) *Eriophyoid Mites, Their Biology, Natural Enemies and Control*. Elsevier, Amsterdam, pp. 611–618.

Thacker, J.R.M. (2002) *An Introduction to Arthropod Pest Control*. Cambridge University Press, Cambridge.

Theunissen, J. and Schelling, G. (1996) Pest and disease management by intercropping: suppression of thrips and rust in leek. *International Journal of Pest Management* 42, 227–234.

Tilman, D. (1999) Global environmental impacts of agricultural expansion: the need for sustainable and efficient practices. *Proceedings of the National Academy of Sciences USA* 96, 5995–6000.

Torchin, M.E., Lafferty, K.D., Dobson, A.P., McKenzie, V.J. and Kuris, A.M. (2003) Introduced species and their missing parasites. *Nature* 421, 628–630.

UN (2004) World population to 2300, a report by the United Nations Department of Economic and Social Affairs Population Division. Available at: http://www.un.org/esa/population/publications/longrange2/WorldPop2300final.pdf (accessed 17 March 2009).

UN (2009) Population Newsletter No. 87. United Nations Department of Economic and Social Affairs, New York. Available at: http://www.un.org/esa/population/publications/popnews/Newsltr_87.pdf (accessed 14 October 2009).

USDA (2001) Pest management practices, 2000 summary. Available at: http://usda.mannlib.cornell.edu/usda/current/PestMana/PestMana-05-30-2001.pdf (accessed 4 May 2010).

van Lenteren, J.C. (2000) A greenhouse without pesticides. *Crop Protection* 19, 375–384.

van Lenteren, J.C. (2006) The area under biological control and IPM in greenhouses is much larger than we thought. *Sting* 29, 78.

van Lenteren, J.C. (ed.) (2008) IOBC Internet Book of Biological Control. Available at: http://www.unipa.it/iobc/download/IOBC%20InternetBookBiCoVersion5 January 2008.pdf (accessed 18 November 2008).

Vila-Aiub, M.M., Vidal, R.A., Balbi, M.C., Gundel, P.E., Trucco, F. and Ghersa, C.M. (2008) Glyphosate-resistant weeds of South America cropping systems: an overview. *Pest Management Science* 64, 366–371.

Vincent, C., Hallman, G., Panneton, B. and Fleurat-Lessard, F. (2003) Management of agricultural insects with physical control methods. *Annual Review of Entomology* 48, 261–281.

Waage, J.K. (1997) Biopesticides at the crossroads: IPM products or chemical clones? In: Evans, H.F. (chair) *Microbial Insecticides: Novelty or Necessity? BCPC Symposium Proceedings No. 68*. British Crop Protection Council, Farnham, UK, pp. 11–20.

Walker, K., Mendelsohn, M., Matten, S., Alphin, M. and Ave, D. (2003) The role of microbial Bt products in US crop protection. In: Metz, M. (ed.) *Bacillus thuringiensis, A Cornerstone of Modern Agriculture*. Haworth Press, Binghamton, New York, pp. 31–51.

Walter, D.E., Krantz, J. and Lindquist, E. (1996) Acari: The Mites. Version 13, December 1996. The Tree of Life Web Project. Available at: http://tolweb.org/Acari (accessed 13 July 2010).

Warabieda, W., Olszak, R.W. and Dyki, B. (1997) Morphological and anatomical characters of apple leaves associated with cultivar susceptibility to spider mite infestation. *Acta Agrobotanica* 50, 53–64.

Way, M.J. and van Emden, H.F. (2000) Integrated pest management in practice – pathways towards successful application. *Crop Protection* 19, 81–103.

WeedScience.com (2010) *International Survey of Herbicide Resistant Weeds*. Available at: http://www.weedscience.org/In.asp (accessed 24 March 2010).

Yaninek, J.S. and Herren, H.R. (1988) Introduction and spread of the cassava green mite, *Mononychellus tanajoa* (Bondar) (Acari: Tetranychidae), an exotic pest in Africa and the search for appropriate control methods: a review. *Bulletin of Entomological Research* 78, 1–13.

3 Pest Management with Biopesticides

There is no internationally agreed, formal definition of a biopesticide. The description favoured in this book – which is one that tends to be followed by many national governments for the purposes of regulation – is that a biopesticide is a mass-produced, biologically based agent manufactured from a living microorganism or a natural product and which is sold for the control of plant pests. The agents used as biopesticides are usually broken down into three categories: (i) microorganisms (bacteria, fungi, oomycetes, viruses and protozoa); (ii) biochemicals (which include plant products such as essential oils, and various compounds synthesized by other organisms such as chitin and chitosan); and (iii) semiochemicals (most of which are insect pheromones used in traps or for mating disruption). The authorities in the USA also include genes that are engineered into crops using recombinant DNA technology to protect them against pests (these genes are known as plant incorporated protectants, PIPs). All of the agents approved for use under these categories of biopesticide are considered by the authorities to present a low risk to people, wildlife and the environment. For this reason, some regulators such as the US Environmental Protection Agency (EPA) group biopesticides together with other low-risk products such as commodity chemicals. If you visit the active ingredient index on the US EPA Biopesticides website (http://www.epa.gov/pesticides/biopesticides/), for example, you will find that true biopesticides are listed alongside inorganic compounds such as kaolin and iron phosphate. While it makes sense to list all these low-risk substances together, it is misleading to label minerals and other compounds that are not made by living organisms as biopesticides.

Worldwide there are around 1400 biopesticide products being sold (Marrone, 2007; R. Gwynn, personal communication). At the time of writing this book (April 2010), there were 68 biopesticide active ingredients registered in the EU and 202 in the USA (plus an additional 16 PIPs in the USA) (Table 3.1). The EU biopesticides consisted of 34 microbials, 11 biochemicals

Table 3.1. Number of biopesticide active substances available in the USA or on the EU Active Substances List (April 2010).

	Insecticide		Fungicide		Herbicide		Nematicide		Attractant		Repellent		Plant growth regulator	
	USA	EU	USA	EU	USA	EU	USA	EU	USA	EU	USA	EU	USA	EU
Microorganisms														
Bacteria	31	4	21	3	–	–	1	–	–	–	–	–	1	–
Fungi	9	5	18	15	5	–	2	1	1	–	–	–	–	–
Viruses	8	5	4	1	–	–	–	–	–	–	–	–	–	–
Oomycetes					1									
Biochemicals	11	3	5	1	1	–	3	–	5	–	23	4	4	3
Semiochemicals	–	–	–	–	–	–	–	–	48	23	–	–	–	–
PIPs														
Bacteria	15	–	–	–	–	–	–	–	–	–	–	–	–	–
Viruses	–	–	1	–	–	–	–	–	–	–	–	–	–	–
Total	74	17	49	20	7	–	6	1	54	23	23	4	5	3

PIP, plant incorporated protectant.

and 23 semiochemicals, while the USA portfolio comprised 102 microbials, 52 biochemicals and 48 semiochemicals. There are about 225 microbial products being sold in total in the countries of the Organisation for Economic Co-operation and Development (OECD) (Kabaluk and Gazdik, 2007). These biopesticide products represent a very small fraction of the total pesticide (conventional chemical plus biopesticide) market. In 2005, global biopesticide sales were put at US$700 million, equivalent to just 2.5% of the total pesticide market (Bailey et al., 2010). However there is rapid growth in the biopesticides sector, with a 5-year compound annual growth rate of 16% compared with 3% for synthetic pesticides and an expected global market worth up to US$10 billion by 2017 (Marrone, 2007). The number of biopesticide active ingredients is also small compared with the number of species of arthropod (predator and parasitoid) natural enemies being used for biologically based pest control. In Europe, about 80 species of arthropod natural enemies are commercially available (EPPO, 2002) while over 850 species have been used worldwide (Cook et al., 1996). However, as mentioned in Chapter 2, biopesticides represent the bulk of the financial market share for biologically based pest control agents (i.e. biopesticides plus predators and parasitoids) at about 85% of the value of this market, compared with approximately 15% for arthropod natural enemies (Gelernter, 2005).

Despite being just a small part of the overall pesticide market, biopesticides encompass a very broad range of living and non-living entities, with varying modes of action and a wide range of characteristics. However, because they are all classed as biopesticides, they are all governed by the same overarching regulation. Usually, this is based on plant protection products legislation that was originally designed for synthetic chemical pesticides, and which may or may not have been modified to take biopesticides into account. This poses challenges for government regulatory agencies in terms of managing the potential human safety and environmental risks of biopesticides, as well as issues of efficacy and the need to deliver government policies that improve the sustainability and productivity of agriculture. For an area such as regulating the potential health risks to farm operatives who are handling biopesticide products, it makes some sense to include microbial control agents and natural products within the same legislative framework as synthetic chemical pesticides because the potential routes of operator exposure are usually identical (e.g. risks of skin contact and inhalation of sprays, dusts or powders in a concentrated form). However, when it comes to environmental risk assessment, the issues affecting natural products are very different from those affecting microbial agents. The potential environmental risks of natural products can reasonably be grouped together with those of synthetic pesticides. In contrast, the environmental risks associated with microbial biopesticides have much more in common with the risks associated with higher organisms used for pest control, such as predators, parasitoids and parasitic nematodes.

For making the best use of a product within the framework of sustainable IPM, then it is undoubtedly the case that the use of the 'chemical pesticide regulatory model' for microbial agents has impeded their development by drawing attention away from their beneficial characteristics as living organisms, in particular their ability to reproduce within host populations and thereby give extended pest control, and instead has forced them to be used as synthetic pesticide clones with unrealistic expectations of chemical-like efficacy (Waage, 1997).

Comparing Microbial Agents with Other Natural Enemies Used in Biological Control

Biological control programmes operate throughout the world in agriculture and forestry. The organisms used for biological control are 'natural enemies': a natural enemy is an organism that kills or debilitates another organism. When a natural enemy is used for biological control, it is referred to as a biological control agent. Biological control agents include the following: (i) predatory insects and mites; (ii) parasitoids, which are insects with a free-living adult stage and a larval stage that is parasitic on another insect; and (iii) parasites, microbial pathogens and antagonists, such as nematodes, fungi, bacteria, viruses and protozoa (Bellows and Fisher, 1999; Bale et al., 2008). Some examples of microbial control agents are given in Table 3.2.

Natural enemies represent a large component of the world's biodiversity. It has been estimated that each agricultural pest species is affected by 50–250 species of natural enemies (van Lenteren, 2000). These organisms play a key role in preventing pests reaching damaging levels. It has been suggested that about 99% of all potential pests are controlled by the natural enemies or antagonists that reside naturally within agro-ecosystems (Thacker, 2002). However, only a small proportion of the available species have been investigated for use as biological control agents. Thus, while approximately 750

Table 3.2. Examples of microorganisms used as control agents of agricultural pests. (For more information see Copping, 2004.)

Organism	Use	Pest	Target crops
Bacteria			
Agrobacterium radiobacter	Antibacterial agent	Crown gall (*Agrobacterium tumefaciens*)	Soft fruit, nuts, vines
Xanthomas campestris pv. *poannua*	Herbicide	Annual bluegrass	Turf
Bacillus subtilis	Fungicide	*Fusarium, Pythium, Rhizoctonia* spp.	Legumes, cereals, cotton
Bacillus thuringiensis	Insecticide	Various Lepidoptera, Diptera, Coleoptera	Vegetables, fruit, cotton, rice, forestry
Fungi			
Lecanicillium longisporum	Insecticide	Aphids	Glasshouse edible and ornamental crops
Phytophthora palmivora	Herbicide	Strangler vine	Citrus
Trichoderma harzianum	Fungicide	*Pythium, Phytophthora, Rhizoctonia* spp.	Orchards, ornamentals, vegetables, glasshouse crops
Protozoa			
Nosema locustae	Insecticide	Grasshoppers, crickets	Pasture
Viruses			
Cydia pomonella granulosis virus	Insecticide	Codling moth	Apple, pear

species of insect pathogenic fungi are known, fewer than 20 have received serious attention as microbial biopesticides of insect pests (Hawksworth *et al.*, 1995).

Before we start to look in detail at microbial biopesticides, it is worth considering for a moment a particular group of biological control agents that sit halfway between the 'macro' agents that include predators and parasitoids and the 'micro' control agents that includes bacteria, viruses, fungi and the like. Pathogenic nematodes are multicellular animals, about 0.5 mm in length, which are used for the biocontrol of insects and molluscs and show close similarities to microbial pathogens in the ways they infect, kill and reproduce in their hosts. Like predators, parasitoids and the authorized microbial biopesticide products, they are recognized as posing minimal or zero safety risk to people. Entomopathogenic nematodes in the genera *Xenorhabdus* and *Heterorhabditis* kill their insect hosts using symbiotic bacteria, which the nematodes carry in specialized internal structures and which are released during infection to cause a fatal septicaemia in the insect host. For these reasons, many biocontrol researchers, practitioners and learned societies, such as the Society for Invertebrate Pathology, tend to group nematodes with the true microbial pathogens, while recognizing the differences. However, in the USA, the EU and other OECD countries their use is exempted from biopesticide regulations. Instead they fall within the legislative framework for the use of higher organisms. These regulations are usually concerned with the introduction of non-native species. For example, in the UK, the use of non-native species of nematodes for biological control is prohibited, being enforced through the UK Wildlife and Countryside Act and overseen by the Department for the Environment, Food and Rural Affairs. In this book, we give some coverage to nematodes for biological control because we believe the comparison with 'true' microbial biocontrol agents is informative. Indeed, from an IPM perspective, it is important to consider microbial biopesticides using the same concepts that are applied to other natural enemies. Because all natural enemies are alive, they share common ecological properties, including the potential to reproduce and the requirement for nutrition and other resources. When applied as a biological control agent, the ability of a natural enemy to multiply and persist within the vicinity of its host or prey enables it to give a degree of self-sustaining pest control. Natural enemy reproduction is an important component of most biological control strategies. Unfortunately, it has tended to be overlooked for microbial biopesticides as a result of them being developed and regulated by a system designed originally for chemical agents.

The Biology of Microbial Natural Enemies

Fungi, bacteria, viruses and protists are all being used for biological control or are being investigated as potential biological control agents. These beneficial microorganisms are widespread in nature, where they contribute to the natural regulation of invertebrates, plant diseases and weeds. They have a range of properties that make them desirable for IPM (Hajek, 2004). The microorganisms selected for use as biological control agents do not naturally infect vertebrates, and so are considered safe to humans, livestock and vertebrate wildlife. They produce little or no toxic residue, and development and registration costs are significantly lower than those of synthetic chemical pesticides (Hajek, 2004). They can be applied to crops using the same equipment used to apply chemical pesticides, formulated in similar ways to pesticides to enhance their efficacy, and they tend to have very short harvest intervals. As living organisms, they can reproduce on or in close vicinity to the target pest, and this provides various degrees of self-perpetuating control depending on the agent used. Microbial agents may also prevent the target pest from feeding or reproducing. Microbial agents used for biocontrol are selective, making their use compatible with the deployment of other natural enemies. The exact host range varies among different types of control agent. Baculoviruses, for example, are highly specific and will generally infect only a single species or genus of insect. In contrast, some species of entomopathogenic fungi are able to infect insects across a range of taxonomic orders (Tanada and Kaya, 1993). Many microbial biopesticides can be produced locally, which can be important in terms of choosing and matching natural enemies to small-scale needs (Waage, 1997). Like other biological control agents, the successful use of microbial natural enemies requires fundamental knowledge of the pathogen and the pest (Collier and van Steenwyck, 2004; Bale *et al.*, 2008). When this condition is satisfied, and the agent is used firmly within IPM, then biological control can sometimes be more cost effective than purely chemical control (Tisdell, 1990; van Driesche and Bellows, 1995). However, under the present market system, many microbial biopesticides products have not competed well with less expensive and more effective synthetic pesticides (Gelernter, 2005). The downsides of using microbial natural enemies are that many are niche-market products, pest control is not immediate, there can be a lack of environmental persistence, and efficacy can be unpredictable in outdoor environments. Microbial agents that can only be cultured *in vivo* can also be expensive to mass-produce. Like all living organisms, they are adapted to certain physical conditions and this can restrict the environments in which they are used. For example, fungi require conditions of high water availability to germinate and grow, and hence fungal control agents naturally function best in the soil or in environments where humidity can be controlled, such as glasshouses. Baculoviruses and entomopathogenic bacteria tend to be susceptible to degradation by ultraviolet light, and so they may not persist well when sprayed on to plant foliage. However, some of these limitations can be overcome using formulations such as stickers, wetters, oils and ultraviolet protectants. Often, efficacy can be improved significantly by paying careful consideration to the method of application, for example by using low-volume or ultra-low-volume sprays and mists that place the microbial agent on the underside of leaves.

Strategies for Using Biological Control Agents, Including Microbial Agents

The ways in which biological control agents are used vary according to the type of pest and the biological characteristics of the control agent, as well as the agricultural setting. Unfortunately, researchers and other specialists working with biological control agents of arthropods, plant pathogens and weeds have adopted different terms to describe the control agents and how they are used. These terms can have subtly different meanings, which can cause confusion. For example, entomologists nearly always refer to insect parasitoids and predators as 'natural enemies', but microbiologists working with pathogens of insects rarely use this term. Those working on the microbial control of plant pathogens often refer to their control agents as 'antagonists', which reflects the fact that many microbial control agents of plant disease work by outcompeting plant pathogens for space or nutrients rather than killing or injuring them. This term is never used by insect pathologists working on insect microbial control, as the microbial agents used for insect control work by directly killing the insect host. The lack of a common language is undoubtedly an impediment for the biological control policy network, including regulators, policy makers, farmers and growers. Some of the terms used have little or no scientific meaning, or they were derived from concepts that were valuable at the time but are now confusing. For example, the use of an alien biological control agent to control an alien, invasive pest is referred to as 'classical' control. The word 'classical' has no inherent biological meaning; rather it refers to the fact that introductions of alien natural enemies to control alien pests were historically the first successful examples of biological control (e.g. the control of cottony cushion scale on citrus in California in the late 1800s) and thus are considered an established model for biocontrol.

Since the beginning of the 21st century, there has been a drive to unify some of the terms used in biological control (Eilenberg et al., 2001). In this book, we are using the Eilenberg et al. (2001) unified definitions of biocontrol strategies as much as we can. There are three broad biocontrol strategies: (i) augmentation (i.e. application of natural enemies that already live in the area of use); (ii) introduction (i.e. release of an alien control agent to control an alien pest); and (iii) conservation (i.e. manipulating agricultural practices or the environment to enhance natural control).

Augmentation biological control

Augmentation is a strategy that uses, as a biological control agent, a species of natural enemy that lives naturally in the country or region of use. Natural enemies are widespread in the environment and exert a degree of control over pests as part of normal ecological processes. The augmentation strategy is aimed at increasing the effectiveness of control by releasing a control agent in crop environments where it is absent or at low levels. The natural enemy is released without the expectation that it will establish permanently or give autonomous regulation of the pest. The word 'augmentation' derives from the idea that individuals of the same species, which have been grown in a specialized production facility, have augmented local natural enemies. 'Augmentation' is used very widely within the specialist biological control literature, particularly for biological control using arthropod natural enemies. However, it is not a self-explanatory term and it almost certainly causes confusion for those outside biological control research.

There are two different approaches to augmentation biological control. The *inundation* approach is 'The use of living organisms to control pests when control is achieved exclusively by the organisms themselves that have been released' (Eilenberg et al., 2001). Individuals of the biological control agent are mass-produced in specialized production plants and then applied in very large numbers into the close vicinity of the target pest. The intention is to achieve rapid pest control, with no expectation that the control agent will reproduce itself in the environment. As a consequence, applications will have to be repeated if the pest population increases again past the economic threshold. Microbial biopesticides used for the inundation approach are usually formulated as liquids, emulsions, powders or granules so that they can be applied using the same kind of sprayers and other apparatus used for the application of chemical pesticides. The inundation approach is based on the hypothesis that natural factors limiting the persistence and spread of the biological control agent (such as abiotic conditions, poor dispersal of the control agent, or a patchy distribution of the target pest) can be overcome by repeated applications of large numbers of the control agent (Hallett, 2005). An example of the inundation approach is the use of the insect pathogenic bacterium *Bacillus thuringiensis*. This bacterium produces an insecticidal protein crystal during sporulation, and a mixture of bacterial spores and the protein crystal is sprayed on to insects for their control. When they are ingested, they cause ion leakage from gut cells and kill the target insect in 24–48 h. However, for most target pest insects, the bacterium does not reproduce to any significant extent and control is caused by the combination of the protein crystals and the individual bacterial cells that have been sprayed.

The *inoculation* approach is 'The intentional release of a living organism as a biological control agent with the expectation that it will multiply and control the pest for an extended period, but not that it will do so permanently' (Eilenberg *et al.*, 2001). Thus, the control agent is able to respond to changes in the size of the pest population. The biological control agent consists of a product that has been grown or cultured in a specialized facility, but smaller numbers are applied to the target pest compared with the inundation approach, and the environment must be suitable to support its reproduction and localized spread. As an example, many of the microbial control agents used against plant pathogens in the soil rely on the ability of the control agent to grow and multiply in the root zone, particularly for those agents that function by outcompeting plant pathogens for space or other resources (Whipps, 2001). The length of time over which the inoculum continues to reproduce in the environment varies according to each situation, and is dependent upon the life history of the control agent (r- or K-selection, specialist or generalist) and the highly complex ecological interactions between the control agent, the pest, the environment, and other components of the biotic community including, of course, the crop plant. In general terms, over time the population of the control agent will be supported at the natural carrying capacity of its environment, because the species of natural enemy that has been released is already a natural component of the biota in the region of use. Successful control relies on the assumption that the control agent is not present in effective numbers within the immediate vicinity of the pest as a consequence of agricultural practice.

In practice, because all living biological control agents have potential to reproduce in the environment, there is not a straightforward distinction between inundation and inoculation. Thus, control agents applied according to the inundation approach will exhibit a certain amount of localized reproduction and spread, resulting in a small amount of self-sustaining control. In inoculation, most control may be given by the released organisms with the effects of the progeny declining over time (Hajek, 2004). For this reason, they are always considered together as forms of augmentation control. All microorganisms used for augmentation control come within our definition of biopesticides, regardless of whether they work by the inundation or the inoculation approach, and their use is covered by biopesticides regulations. However, the regulatory situation for the next strategy for biological control, classical control, is not so clear-cut.

Classical biological control

Classical biological control is 'the intentional introduction of an exotic biological control agent for permanent establishment and long-term pest control' (Eilenberg *et al.*, 2001). In this context the term 'exotic' refers to a non-indigenous organism. It is based on the natural enemy release hypothesis and is targeted at the control of alien invasive species that have established and spread in a new country as a result of separation from their co-evolved natural enemies in their homeland. It is intended to work over very large geographical scales. A limited number of releases of the biological control agent are made, with the expectation that a permanent relationship between pest and its co-evolved natural enemy will become re-established, i.e. invasion by the natural enemy is desired. In order for it to work successfully, in other words to permanently control the pest population with no adverse side effects of the biological control agent on the environment, then a detailed and thorough ecological knowledge is required. Classical biological control is not commercially profitable, and for this reason it is implemented through government-funded programmes. Invasive alien species cause very significant amounts of economic and ecological damage and classical control is the only realistic option available for permanent control in many cases. It has been used successfully in the past, although it is true to say that, in Europe at least, introductions have become more difficult in recent years because of public fears over environmental impact, an absence of appropriate legislation and some high-profile cases where unsuitable, generalist natural enemies have been released due to ineffective regulations and impacted beyond their intended target. A more detailed discussion of the environmental safety issues surrounding classical control is included at the end of the chapter.

The first, and still the best known, example of classical control concerned the cottony cushion scale, *Icerya purchasi*. This insect was unintentionally introduced into Californian citrus orchards from Australia in 1868 and pushed the Californian citrus industry to the verge of collapse in less than 20 years. Natural enemies of the scale were identified in Australia and around 500 predatory vedalia beetles, *Rodolia cardinalis*, were released. The beetle population expanded rapidly and their subsequent spread was aided by moving beetle-infested branches to new orchards. The strategy was so successful that the scale was completely controlled by 1890. Today, classical biological control of insect pests is used on 350 million ha worldwide, equivalent to 10% of the total global area of cultivated land, and is reported to have a cost–benefit ratio in the range of 1:20 to 1:500 (Bale *et al.*, 2008). Most of the natural enemy species introduced have been insects: around 2000 species have been introduced worldwide, leading to the permanent reduction of 165 arthropod pest species (van Lenteren *et al.*, 2006). There are estimated to have been about 130 examples of classical introductions against insect pests in Europe (Sheppard *et al.*, 2006). Microbial natural enemies have been used less frequently in classical control. Around 135 classical introductions of microbial pathogens including nematodes have been recorded for the control of insect and mite pests (Hajek and Delalibera, 2010). We will include some examples as we go through this chapter.

Classical release of microbial natural enemies for control of crop pests does not fall within our definition of a biopesticide. Classical microbial control agents are not mass-produced entities and they are not sold for commercial profit. However, we will consider them briefly in this book for two reasons. First, knowledge of classical control is very instructive when considering the potential environmental risks associated with microbial biopesticides. Second, classical microbial control agents are affected by plant protection products legislation. Unfortunately, as is the case for microbial biopesticides, the plant protection products legislation in place in organizations such as the EU was not developed with biological control in mind and this is hindering the development and use of microorganisms in classical control (Sheppard *et al.*, 2006). For example, the EU plant protection products legislation requires information on directions of use of the product to be available to users and detailed information on field efficacy in comparison with standard chemical treatments, neither of which is appropriate to classical control. Registration costs, including the costs of statutory toxicological testing, are also very high for these agents, which are developed using the limited funds that are available from public money.

Conservation biological control

Conservation biological control is 'modification of the environment or existing practices to protect and enhance specific natural enemies or other organisms to reduce the effect of pests' (Eilenberg *et al.*, 2001). It does not require application of natural enemies, but is intended to enhance the numbers and activities of natural enemies living naturally within the vicinity of pest populations. Like other forms of biologically based control, it is not a stand-alone solution and must be firmly embedded within the overall context of IPM. Conservation biological control techniques are in place for enhancing the activities of arthropod natural enemies and there is a large body of evidence suggesting that provision of floral resources, alternative food and shelter habitats can increase the abundance and diversity of predators and parasitoids, but there is as yet limited evidence that this leads to decreased pest damage or increased yields (Jonsson *et al.*, 2008). However, successful conservation biocontrol of insect pests with predators and parasitoids has been demonstrated for a small number of crops including outdoor lettuce and grapes (Chaney, 1998; Berndt and Wratten, 2005; Berndt *et al.*, 2006). Because it entails habitat management, it could provide additional public goods, such as biodiversity conservation, ecological restoration and tourism (Fielder *et al.*, 2008). Wine producers in New Zealand, for example, are now growing flowers in between vines in order to lure and retain naturally occurring parasitoids; these floral displays also attract tourists to the region and are being used to market the wine as a premium, environmentally friendly brand (Anon., 2006).

Conservation biological control can also take advantage of microbial natural enemies. Successful examples include: (i) the exploitation of microorganisms that antagonize plant pathogens in disease-suppressive soils; (ii) the development of a service that predicts natural outbreaks of the insect pathogenic fungus *Neozygites fresenii* in populations of cotton aphids in the south-eastern USA; and (iii) reducing or avoiding fungicide sprays that inhibit the entomopathogenic fungus *Neozygites floridana*, which improves the natural biological control effect of this fungus against the two-spotted mite, *Tetranychus urticae*, in the maize–groundnut–soybean agrosystem in the USA (Smitley *et al.*, 1986; Dick *et al.*, 1992).

Using a microbial natural enemy for conservation biological control is patently different from application of a microbial biopesticide: it requires no product to speak of and falls outside plant protection products legislation; hence there is no statutory testing of safety or efficacy. It is based on an ecological hypothesis that manipulating agricultural practice can increase the activity of microbial natural enemies that are residing within the locality. It is important to keep this in mind when considering the principles of environmental risk evaluation for microbial biopesticides, not least because augmentation microbial control and conservation microbial control use the same kinds of microbial natural enemy. The main difference concerns the level of exposure to microorganisms in terms of human safety. Because augmentation microbial control uses mass-produced microorganisms, there are potential safety risks associated with sensitization and developing allergies for people who regularly come into direct contact with high concentrations of microorganisms, particularly if there is a chance that microbial spores and other entities could be inhaled. For classical control using a limited number of small releases, the exposure is likely to be confined to just the very small number of people doing the releases.

Microbial Natural Enemies of Invertebrate Pests, Plant Pathogens and Weeds

Pathogens of invertebrates and their use in biological pest control

Invertebrates are utilized as a source of food and for reproduction by a wide variety of microbial parasites in nature. Those pathogens with suitable properties – in terms of their ecological characteristics, safety to humans and the environment, and the ease with which they can be used in a practical way – can make useful biological control agents. The pathogens of insect and mites are collectively referred to as entomopathogens. There are also pathogens of molluscs that can be used for biocontrol.

The main groups of invertebrate pathogens used for biocontrol are fungi, bacteria and viruses. The entomopathogens used for biological control all work by causing lethal infections in their hosts, although their modes of action vary significantly depending upon the taxonomic group of pathogen. For this reason, it is sensible to divide up our discussion of the biology of entomopathogens according to their taxonomic classification. In contrast, the biggest division between the microbial control agents of plant disease is functional rather than taxonomic. Here, there is a logical division between microbial antagonists, which interact directly with plant pathogens, and agents that act indirectly by boosting a plant's ability to withstand disease.

Insect pathogenic bacteria
Entomopathogenic bacteria occur in about five different taxonomic families. Most of these are opportunistic pathogens that only infect insects that are under some kind of environmental stress. In contrast, a small number of bacterial species have evolved as specialist insect pathogens. The most important species occur in the genus *Bacillus*. Two species of these Gram-positive, spore-forming bacteria are used as augmentation biological control agents using the inundation approach. *Bacillus sphaericus* is a pathogen of mosquito larvae and has been used for their control (Lacey and Undeen, 1986). *Bacillus thuringiensis* (known simply as Bt) is naturally widespread and can be found in soil, on plant surfaces and in grain storage dust. It consists of a range of different subspecies and strains, each of which is pathogenic to a narrow range of insect species. As part of its life cycle, Bt produces a large protein crystal associated with spore formation, known as a parasporal crystal. There is normally one crystal per cell. The proteins, referred to as δ-endotoxins, are formed outside the spore and are toxic to insects, although some types of Bt produce crystals with no known activity. The δ-endotoxin is host-specific, does not harm vertebrates and is very active (Bond *et al.*, 1971; Siegel, 2001). Wild-type strains of Bt usually have more than one crystal gene, called *cry* genes, which occur on plasmids (extrachromosomal DNAs) and the parasporal crystal usually consists of combinations of different proteins. Following ingestion, the alkaline environment of the insect midgut causes the crystals to dissolve. At this stage, the toxins are inactive, and are trimmed by gut proteases to an activated, N-terminal, truncated form. The activated toxin binds to specific receptors on the midgut epithelial cells, inserts itself into the membrane using the N-terminus of the protein and forms pores that kill the epithelial cells by osmotic lysis (Gill *et al.*, 1992). In some highly susceptible hosts, ingestion is followed by a general paralysis leading to death within an hour. In the majority of insects, only the gut is paralysed and death occurs in about 48 h depending on dose.

Bt δ-endotoxins are part of a large family of homologous proteins. More than 130 genes encoding δ-endotoxins have been identified to date. Differences in the nucleotide sequence of the *cry* genes cause variation in the toxicities of different crystals. These genes generate a rich source of biodiversity and partly account for the host preferences expressed by different subspecies of Bt. Bt is the most widely used microbial control agent and it is used always as an inundative biopesticide. The bacterium is grown in fermentation tanks up to the sporulation phase, after which the spores and the δ-endotoxin crystals are harvested, cleaned, dried and formulated. The final product is then sold to growers and farmers. At the moment the available products use toxins specific for caterpillars, beetles or haematophagous flies but new toxins are being discovered against a wider array of targets including mites, cockroaches, grasshoppers and other species (van Frankenhuyzen, 2009). Bt is very well documented as being safe to workers, beneficial organisms and the environment (Lacey and Siegel, 2000). After production, the bacterial spores and crystals are processed to improve their shelf-life and handling characteristics. They can be formulated for application as dusts, granules, wettable powders, or as liquid emulsions, encapsulations or suspension concentrates. Booster formulations are also available using semiochemical attractants or phagostimulants (Brar *et al.*, 2006).

At present, microbial Bt formulations are applied as foliar sprays for fruit and vegetable crops (they have also been very important for the control of insect vectors of human and animal disease including mosquitoes, blackflies and tsetse flies). Prior to the mid-1990s, Bt sprayable products were also sold for use on broad-acre crops; however, these have now been superseded by the development of Bt transgenic crops (see below). Nevertheless, Bt sprays remain an important and growing tactic for horticultural crops where their high level of selectivity and safety are considered desirable, and where there are problems with resistance to conventional insecticides, such as apple production. In the USA, for example, Bt foliar sprays are applied to about 10% of the total area of cultivated apple trees. By comparison, the chemical insecticide chlorpyrifos (an OP) is applied to 60% of the total area, imidacloprid (a neonicotinoid) to 25% and Spinosad (a microbial) to 24% (NASS, 2008). Bt foliar sprays cost about the same as 'modern' chemical pesticides such as Spinosad (van Driesche *et al.*, 2008).

Expression of Bt δ-endotoxins in genetically modified crops
While Bt sprays can be very effective products, their use has been eclipsed by the development of GM crops that express Bt *cry* genes. These crops are commonly referred to as Bt crops and they have been adopted rapidly by farmers around the world, with the exception of Europe where there has been widespread reservation among consumers about GM crop technology. In 2009, GM crops were being grown in 25 countries over 134 million ha, equivalent to about 9% of the global area of cultivated land (International Service for the Acquisition of Agri-Biotech Applications, 2010). These include six EU states – Spain, Czech Republic, Portugal, Poland, Slovakia and Romania – although the area of cultivation in these countries is generally less than 0.1 million ha.

Bt GM crops were developed to overcome some of the perceived shortcomings of using Bt sprays, specifically the lack of systemic activity (which meant that Bt sprays were not effective against insects that burrowed into plant tissue), degradation by ultraviolet light and too narrow a host range (de Maagd *et al.*, 1999). The crops that are currently available express *cry* genes for δ-endotoxins that are specific for lepidopteran pests, particularly those that are major pests of maize, such as European corn borer, *Ostrinia nubilalis*, and pests of cotton such as the bollworm, *Helicoverpa zea*. Bt maize with resistance to western corn rootworm, *Diabtrotica virgifera* (Coleoptera), which is a highly damaging and invasive pest of maize in North America and Europe, is also available, while the technology has been recommended as the way forward for managing diamondback moth, *Plutella xylostella*, which is a particularly severe pest of brassica production in Africa and South Asia (Grzywacz *et al.*, 2010). Truncated forms of Bt *cry* genes were expressed in plants following *Agrobacterium*-mediated transformation. The nucleotide composition of bacterial and plant genes varies significantly. As a consequence, the original *cry* sequences inserted into plants contained regions that acted in the plant as mRNA instability motifs, as well as regions that were interpreted by the plant as transcription termination sites, which meant that expression levels were very low. In total, nearly 60% of the codons of the inserted gene were altered, which increased expression levels 100-fold and gave effective protection (de Maagd *et al.*, 1999).

Since being made commercially available in the late 1990s to early 2000s, Bt maize and cotton have become mainstays in many countries and are grown on tens of millions of hectares. Surveys of farmers' pesticide use indicate that growing Bt crops can result in significantly reduced spray applications of conventional insecticides, up to 70% in some cases (Shelton *et al.*, 2002.). Because δ-endotoxins are highly specific, this can result in significant environmental benefits. Laboratory and field studies done over multiple years indicate that Bt crops have substantially lower impact on non-target organisms than the broad-spectrum insecticides used previously, and have shown either no impact or only a transient effect on arthropod natural enemies (where a reduction has occurred, it has been caused by a reduction in the number of target pests as prey for natural enemies and not by a direct effect of the crop) (Romeis *et al.*, 2006).

Despite this, there has been a very restricted uptake of Bt crops in Europe. Why is this? It is undoubtedly the case that the development of policies on GM crops in Europe has been affected by a lack of 'upstream' engagement between governments, regulators, farmers, pressure groups, industry, the media and other members of civil society. In addition there has been a loss of confidence in scientific experts by the general public over issues of risk. Governments have based much of their policy on evidence-based risk assessment, forgetting perhaps that while scientific expertise and evidence can help answer specific questions about GM crops, it cannot be the only tool for developing policy. People entering the debate about GM have different points of view and hence a resolution may not be possible. Agriculture in Europe has its own distinctive social dimension and hence Europeans may well have different concerns about GM compared with citizens elsewhere. The ethical issues surrounding GM crops are many and complex, and include general welfare (i.e. the responsibility of governments to protect the interests of citizens), consumer choice and rights, principles of justice, and the boundary between what is considered natural/unnatural (Nuffield Council on Bioethics, 1999).

Those in favour of GM argue that transgenic crops can help increase yields (e.g. by reducing losses due to pests), can have improved nutritional content, require fewer inputs and have less postharvest spoilage and wastage. Arguments against GM crops include possible harm to human health, concerns that the technology consolidates the industrialization of agriculture, that it is not natural, and that it may damage the environment (e.g. by having effects on natural enemies and other non-target species or by the introgression of transgenes into wild plants). In Europe and elsewhere, detailed environmental risk analysis of potential effects of GM crops, based on laboratory and field experiments, is made before licences to release the technology are granted, and extensive farm-scale evaluations have been done (Freckleton *et al.*, 2003). Critics have argued that even farm-scale trials cannot predict the effects of GM crops when grown at very large scales. However, eight countries now grow more than 1 million ha of GM crops each (the USA, Canada, China, India, South Africa, Paraguay, Argentina and Brazil) (*Science* online, 2010) and hence if negative effects do occur from these crops, then one might assume that they will become apparent in the next few years if they have not done so already. For Bt crops, resistance management is a legitimate concern, as resistance has developed to Bt foliar sprays (Li *et al.*, 2007). A resistance management strategy has been devised based on the cultivation of areas of non-Bt crops as refugia to maintain susceptible alleles within the pest populations, and this has proved successful so far (Shelton *et al.*, 2002). The movement of Bt transgenes into landrace and wild populations of plants is also an important area. This is not an issue in countries where the GM crop species is not native and does not hybridize with native plants (e.g. Bt maize in Europe), but it could be a different story where GM varieties are grown in areas of the world that are centres of plant origin and harbour much of the natural genetic variation in wild relatives of crop plants. The concern is that transgenes, such as those conferring resistance to insect herbivores, could increase the fitness of wild relatives and thus would enhance their survival. Determining the biological significance of the dispersal of transgenes into wild relatives or landraces is extremely challenging, because it requires a detailed understanding of plant population dynamics over large spatial scales and the information required cannot necessarily be obtained from small plot experiments. In Mexico, which represents much of the natural diversity of maize, the cultivation of GM maize varieties has not been allowed since 1998. Nevertheless, a comprehensive recent study indicates that 3% of seed lots of landrace maize in Mexico could express recombinant proteins from Bt while 2% could express recombinant proteins for glyphosate tolerance (Dyer *et al.*, 2009). Transgene dispersal has probably occurred by growing GM varieties that then hybridize with landraces, and hybrid seed is passed from farmer to farmer as part of everyday business. The GM varieties may have been planted prior to the moratorium in 1998, or illegally planted by farmers, or they may have come from GM grain sent from the USA to Mexico intended for food use.

If we set the issue of unintended transgene dispersal aside, it is apparent that Bt GM crops can provide benefits for farmers, but we want to stress that they are not a panacea for insect control, and if they are not used wisely then side effects can and will occur. A good illustration of this concerns secondary pest problems caused by mirid bugs (hemipteran pests with piercing mouthparts that feed on plant sap), which have occurred on Bt cotton grown in China (Lu *et al.*, 2008). Broad-spectrum pesticides controlled the mirids previously, but they are not controlled by Bt cotton, which is specific for caterpillar pests. Problems with mirids in China did not occur until a few years after the widespread uptake of Bt cotton (i.e. there was a time lag between adoption of Bt cotton and the onset of secondary pest problems). The unfortunate result is that some farmers growing Bt cotton in China are having to make more pesticide applications than before in order to control mirid outbreaks, with a net reduction in revenue compared with conventional cotton (Wang *et al.*, 2006). Secondary pest outbreaks are a well-known phenomenon in agriculture, but without more evidence it is difficult to say whether this particular problem could have been foreseen. There is no evidence that the secondary pest outbreaks were caused by the Bt crops having a negative effect on mirid natural enemies within the agro-ecosystem; rather it is because broad-spectrum pesticides previously controlled these pests.

One lesson stands out clearly: if GM crops, or any other new technologies, are to be used to increase crop production, then they must be treated on a case-by-case basis and utilized according to basic IPM and ecological principles in order to make their use sustainable and environmentally acceptable.

Insect pathogenic viruses
Insect pathogenic viruses are widespread in the natural world: over 1600 different viruses have been recorded from more than 1100 species of insects (Tanada and Kaya, 1993). One group of insect pathogenic viruses, the baculoviruses, have received particular attention as potential biological control agents. The viruses in this group only cause infections in insects and other closely related arthropods, and are biochemically and genetically very distinct from any viruses recorded from vertebrates; thus they are considered inherently safe to people. All baculoviruses consist of the same basic structure: an enveloped, rod-shaped nucleocapsid comprised of DNA and protein, surrounded by a proteinaceous occlusion body that confers some environmental protection to the virus. There are two baculovirus genera: (i) nucleopolyhedroviruses (NPVs) infect over 400 insect species, mainly in Lepidoptera (34 families) and to a lesser extent Hymenoptera and some other orders (Diptera, Coleoptera, Neuroptera); and (ii) granuloviruses (GVs) infect lepidopteran hosts, each GV species usually being specific for a particular species of host.

Baculoviruses usually infect larval hosts that acquire virus particles during feeding. The virus infects the cells of the midgut and then spreads to the rest of the body. During infection the host becomes debilitated, resulting in reduction of development, feeding and mobility and increasing exposure to predation.

Death occurs in 5–8 days depending on the amount of virus inoculum that has been acquired, but it can be longer. Diseased and dead larvae serve as inoculum for virus transmission. Virus particles can be spread by rain or by other animals, including other insects.

Because they have such a narrow host range, which is well characterized and documented, baculoviruses are considered to pose minimal environmental risk when used in IPM (Huber, 1986; Groner, 1990). Baculoviruses require living insect cells in order to reproduce and hence mass production can only be done by growing the virus in host insects. This might appear to be an expensive and cumbersome system, but in fact they can be mass-produced economically in the larger lepidopteran species and some very efficient factory mass-production systems have been developed. The harvested viruses can also be formulated to improve their performance in the field, and they are usually applied as a spray on to plant foliage using conventional spray apparatus.

In the USA, baculovirus products are available as inoculative biopesticides for the control of forest pests such as Douglas fir tussock moth and gypsy moth, and as inundative biopesticides against pests of cotton such as *Helicoverpa* and *Spodoptera*. In the USA and Europe the *Cydia pomonella* granulovirus (CpGV) is used as an inundative biopesticide against codling moth on apples. In Washington State, the USA's biggest apple producer, it is used on 13% of the apple acreage (NASS, 2008). In Brazil, the NPV of *Anticarsia gemmatalis*, the soybean caterpillar, was used in the mid-1990s on up to 4 million ha (approximately 35%) of the soybean crop (Moscardi, 1999). The costs of using this virus biopesticide are reportedly 20–30% lower than using chemical pesticides; however, the technical challenges of mass-producing the virus meant that supply could not meet with demand, and as a result the Brazilian government has invested in new production technologies (Szewczyk *et al.*, 2006). It remains to be seen whether use of the virus is supplanted by the imminent commercialization of Bt transgenic soybean (MacPherson and MacRae, 2009). Elsewhere, the non-occluded baculovirus of the rhinoceros beetle, *Oryctes rhinoceros*, has proved to be a valuable inoculative biopesticide to protect coconut palms in the Pacific islands (Huger, 2005). An early example of serendipitous classical control involving a baculovirus concerns the European spruce sawfly, *Gilpinia hercyniae* (Moscardi, 1999). This hymenopteran insect was introduced into North America accidentally from Europe during the early 1900s and occurred in outbreak proportions. It became a serious problem in Canadian forests in the 1930s. A major programme was begun in which predators and parasitoids were imported from Europe and mass-reared for release into the forests. Initially the level of release was low but expanded rapidly (2.5 million parasites released in 1935, 18 million in 1936, 47 million in 1937 and 22 million in 1940). Meanwhile the sawfly problem continued to get worse; in 1939, 73% of white spruce and 43% of black spruce in the original outbreak area were reported dead. By 1942, however, the larval population in all but a few areas had declined to medium or light intensity and this was associated with a viral disease caused by an insect pathogenic virus. The origin of the virus has never been traced but it was probably introduced unknowingly with one of the parasitoids from Europe. Once introduced, the virus spread rapidly and, along with other biological control agents, it kept the sawfly population below the economic threshold level. The success of this virus as a long-term control agent is due to several factors. Transovum transmission (i.e. vertical transmission) of the virus and contamination of foliage with virus by adult sawflies have been postulated as effective mechanisms of spread and transmission.

Entomopathogenic fungi
Most species of entomopathogenic fungi used for biological control occur in two phyla, the *Zygomycota* and the *Ascomycota*. In the *Zygomycota*, most entomopathogenic species occur within the class *Entomophthorales*. These fungi show asexual reproduction on individual hosts during the summer but often switch to sexual reproduction at the end of the year to produce resting spores that survive over winter. The *Ascomycota* contain fungi that show sexual reproduction as well as species that appear to have lost the ability to reproduce sexually.

Entomopathogenic fungi attack all major groups of insects and mites. They infect their hosts using spores that adhere to the outside of the host's body and germinate in response to biochemical and environmental cues. The spores then grow through the cuticle using a combination of enzymatic action and mechanical pressure. Usually the fungus grows into the host's haemocoel, the intercellular space that surrounds the major organs and is bathed in a nutrient-rich fluid. If the fungus is able to overcome the host immune system operating within the haemocoel, it then goes on to proliferate and spread throughout the rest of the body, often in the form of yeast-like cells that do not naturally survive outside the host. The host is killed by a combination of mechanical damage, nutrient exhaustion and cell death caused by fungal toxins. The relative importance of these mechanisms varies with the specific fungal isolate or host. In many cases, a reduction of feeding is one of the first overt changes in an infected host. Death occurs within 4–7 days of infection, followed by the production of large numbers of spores on the cadaver. Thousands or millions of spores may be produced on large insects, such as locusts or caterpillars. The spores of many species of *Entomophthorales* are actively discharged from the cadaver in order to transmit the fungus to new hosts. They also show a range of other adaptations to increase transmission including timing the release of spores to periods of the day that are most favourable to infection and manipulating host behaviour so that diseased insects die in exposed positions (Roy *et al.*, 2006). The sexual *Ascomycetes* show similar high-level adaptations to the life cycles of particular host insects; however, the asexual *Ascomycetes* appear to be more opportunistic.

The *Entomophthorales* contain some species that cause natural epizootics in a range of agricultural pests, although many of these species cannot be grown readily *in vitro* and so they are not yet used widely for commercial biological control. However, these epizootics often reduce the pest below the economic threshold (Lacey and Kaya, 2006) and some systems have been developed to exploit them for crop protection. The best example concerns epizootics of cotton aphids, *Aphis gossypii*, by *N. fresenii* in the southern USA. This fungus causes epizootics that occur at the same times in summer over large areas and which reduce aphid populations by 80% within 3–4 days. By using a simple sampling method, the fungal outbreaks can be predicted up to 10 days in advance and this gives the farmer the option to withhold insecticide sprays. A *Neozygites* sampling and advice service is now run for cotton farmers in ten US states and saves millions of dollars in reduced spray costs every year (Hollingsworth *et al.*, 1995; Pell *et al.*, 2010).

Entomopathogenic fungi in the asexual *Ascomycetes* are associated less commonly with natural epizootics, but they are popular choices for use as inundative biopesticides because they can be mass-produced easily and applied using the same equipment that is used to apply conventional chemical pesticides. Since the 1960s, at least 170 different fungal bioinsecticides and bioacaricides have been developed (Faria and Wraight, 2007). About three-quarters of these products are considered to be currently 'active' (commercially available, registered or undergoing registration). About 50% have been developed in South and Central America, 20% in North America, 12% each in Europe and Asia, and 3% in Africa (Faria and Wraight, 2007). These products are used against a wide range of target pests including species within the Hemiptera, Coleoptera, Lepidoptera, Diptera, Orthoptera and Acari. The majority of products are based on the ascomycetes *Beauveria bassiana* or *Metarhizium anisopliae*, but mycopesticides are also available using *Beauveria brongniartii* and the ascomycete genera *Lecanicillium*, *Isaria* and others. They have been used successfully as control agents of insect and mite pests in glasshouse crops, fruit and field vegetables. The largest single area of use is in Brazil, where commercial biopesticides based on *M. anisopliae* are used successfully against spittlebugs *Mahanarva posticata* and *Mahanarva fimbriolata*, which are major pests of sugarcane. Approximately 750,000 ha of sugarcane were estimated to be treated with *M. anisopliae* in 2008, while the fungus was also applied to 250,000 ha of grassland to control pasture spittlebugs (*Brachiaria* spp.) (Li *et al.*, 2010). The fungus is mass-produced on rice grains and is sold to farmers as fungus-colonized substrate or as a purified spore powder. The cost of using the fungus is put at US$12–60/ha, depending on the level of pest damage. Part of the reason for the success of *Metarhizium* biopesticides is attributed to government support for research and development together with the availability of low-cost substrates for mass production, which keeps the price of the fungal biopesticide competitive with that of chemical insecticides (Li *et al.*, 2010). Different strains of *M. anisopliae* have also been developed commercially for the control of locust and grasshopper pests in Africa and Australia, including the desert locust *Schistocerca gregaria*, which can cause vast amounts of devastation during outbreak years. These pests are usually treated with conventional insecticides sprayed over very large areas, particularly OPs, which raises concerns about environmental and human safety. When mass-produced *Metarhizium* spores are sprayed in an oil-based formulation they cause up to 90% locust and grasshopper control in 14–20 days. The fungus is specific for locusts and grasshoppers and does not affect non-target organisms. Reasons for its effectiveness include not only high virulence but also persistence of fungal spores on the soil and vegetation and the production of new spores on infected cadavers, which then spread to uninfected insects. The fungus is approved for use in several African countries and Australia (Lomer *et al.*, 2001) and the Food and Agriculture Organization (FAO) has recommended that strategies are developed to bring it into use widely in desert locust control programmes (FAO, 2007).

Sixty-seven introductions of entomopathogenic fungi have been made for classical biological control, using 20 fungal species against a range of invasive insects (Thysanoptera, Hemiptera, Orthoptera, Coleoptera, Diptera, Lepidoptera) and Acari (Hajek and Delalibera, 2010). The success rate for establishment of these introductions is estimated at around 60%. There have been no documented cases where classical introduction of an entomopathogenic fungus has caused negative effects to the environment including significant mortality of non-target species (Hajek *et al.*, 2007).

We will mention two examples of classical control with entomopathogenic fungi here. First, the gypsy moth, *Lymantria dispar*, was introduced to the USA from France in the 19th century (*c*.1869) and has caused highly significant damage to the forest ecosystem in the north-east USA, particularly to oaks. *Entomophaga maimaiga*, an entomophthoralean fungus that is specific to gypsy moth, was introduced from Japan in 1910, but it did not establish (Hajek, 2004). However, outbreaks of the same species of fungus started to be observed in 1989. It is not known if these outbreaks are caused by a different strain of the fungus that was introduced by an unknown mechanism, or whether they are caused by the strain released in 1910 and took nearly 80 years to establish. The fungus spreads using infective spores, which are actively discharged from infected larval cadavers. The fungus has caused repeated outbreaks since 1989 and is spreading, but its effects on gypsy moth will only be gauged following long-term monitoring. Second, cassava (*Manihot esculenta*) is the fourth most important carbohydrate source in the tropics after rice, sugar and maize. Cassava is native to South America, but is now grown widely in Africa and Asia. Two major pests were introduced to Africa from South America: the cassava green mite, *Mononychellus tanajoa*, and the cassava mealybug, *Phenacoccus manihoti*, causing yield losses of up to 85% (Bellotti *et al.*, 1999). Application of synthetic pesticides against these pests is not an option for low-income African farmers, but classical biological control has proved successful. The phytoseiid predatory mite *Typhlodromalus aripo*, sourced from Brazil, has proved successful against cassava green mite, while cassava mealybug is being managed using the parasitoid *Apoanagyrus lopezi*. *Neozygites floridana*, an entomophthoralean fungus that causes natural epizootics in populations of cassava green mite in Brazil, is also being investigated as a potential biological control agent. Pathotypes of *Neozygites* from South America, which suppress populations of *M. tanajoa* in areas of high rainfall, are being assessed for release in Africa where native strains of *N. floridana* do not cause regular epizootics (Delalibera *et al.*, 1992; Hountondji *et al.*, 2007).

Pathogenic nematodes

As stated earlier, pathogenic nematodes are not classed as 'true' biopesticides by regulators, although they are used in identical ways to microbial biopesticides. For this reason we shall mention them now briefly. The nematode families Steinernematidae and Heterorhabditidae (commonly termed entomopathogenic nematodes or EPNs) are highly virulent, obligate parasites of insects and are the most important nematodes for biological pest control. The infective stage is the third-stage nematode, known as a 'dauer' juvenile (*Dauer* is German for durability). Heterorhabditid nematodes are hermaphrodites, and therefore only one dauer juvenile is needed to enter the host for progeny production. In contrast, steinernematid nematodes are either male or female and a member of each sex must infect a host for reproduction to occur. Host finding by infective juveniles can be an active process in response to physical and chemical cues. Dauer juveniles of steinernematid species infect their hosts by being ingested or enter through the spiracles and penetrate the tracheae, whereas the heterorhabditids are able to enter a host by actively burrowing through the cuticle. The EPNs introduce symbiotic, pathogenic bacteria of the genera *Xenorhabdus* (in the Steinernematidae) or *Photorhabdus* (Heterorhabditidae) into the haemocoel of their hosts following penetration. Subsequent multiplication of the bacteria leads to host death, which can occur within as little as 48 h of infection. Insects infected with nematodes often have abnormal behaviour compared with uninfected individuals. The nematode can kill its host without its associated bacterium but is unable to reproduce without it. After the host has died, the dauer juvenile nematodes mature into adults and may undergo successive rounds of reproduction depending on the size of the host, but ultimately the infection cycle terminates with the production of large numbers of progeny dauer juveniles. If adequate moisture is present, the next generation of dauer juveniles leave their hosts through rents in the cuticle. In the laboratory, they exit from the host 8–14 days after infection. The dauer juveniles can remain active for a long period of time and, under laboratory conditions, some will live for 1 year (Koppenhofer, 2006).

Infection by EPNs requires the presence of free water. Nematodes show many adaptations to the soil environment and hence they tend to be used for biocontrol of soil-dwelling pests. They can be mass-produced in liquid substrates on an industrial scale, formulated to improve storage and application characteristics. They are usually applied to the soil as a drench but may also be added to crop irrigation pipes. Up to ten commercial EPN inundative biopesticide products have been available in the UK and over 60 have been available in Europe, mainly targeted against pestiferous Coleoptera, Lepidoptera and Diptera on horticultural crops. For example, a number of products are sold by different companies based on *Steinernema feltiae* and used for the control of larvae of the black vine weevil, *Otiorynchus sulcatus*, a serious pest of soft fruit and ornamental crops. They have also been used with commercial success against citrus root weevil, against turf pests and on mushroom crops (Georgis *et al.*, 2006), and have even been shown to give some control against leaf miners when sprayed on to plant foliage (Hara *et al.*, 1993).

Phasmarhabditis hermaphrodita is a rhabditid nematode that is a virulent parasite of a range of terrestrial molluscs, including many key pest species. Its life cycle is very similar to that of the EPNs, although its association with symbiotic bacteria is uncertain. It is mass-produced in the same way as EPNs and is sold as a formulated product for control of slugs and snails. The product is comprised of infective juvenile nematodes mixed with clay to make a water-soluble formulation that kills the host in 4–21 days depending on the dose and environmental conditions. It is used in commercial horticulture to help in slug management on field vegetable crops and is also sold to amateur gardeners. It is currently available commercially in 14 countries in Europe. It has been shown to be native to Chile, but it has not yet been confirmed as occurring naturally in the USA or Canada and so it is prohibited from sale in these countries (Rae *et al.*, 2007).

Microbial control of plant pathogens

Plant pathogens are controlled naturally to some degree by a range of microorganisms, including fungi, bacteria and viruses. Some of these are being used for biological control using augmentation, classical and conservation strategies. The target plant pathogens include fungi, oomycetes, bacteria, viruses and plant parasitic nematodes. The development of microbial biopesticides of plant pathogens is being driven by the increasing withdrawal of synthetic fungicides following government reviews of their safety, but equally important is the worldwide ban on the use of methyl bromide, which was used widely as a soil sterilant but is being withdrawn because it contributes to depletion of ozone in the atmosphere. The commercialization of microbial biopesticides as control agents of plant pathogens and plant parasitic nematodes is a relatively young endeavour; effective products for disease control have only become commercially available to any extent since the mid-1990s (Whipps and Davies, 2000). In 2000, around 80 products were on sale or close to market (Whipps and Davies, 2000).

The microorganisms exploited for plant disease biocontrol have a wide range of modes of action. There are two broad classes. Microbial antagonists occupy the same ecological niche as the target plant pathogen and interact directly with it. The mechanisms of interaction include parasitism, competition for space, water or food, or 'chemical warfare' using antibiotics or other secondary metabolites that harm the target pathogen. The second class involves an indirect effect in which the control agent induces a resistance response in the plant that gives it protection against virulent plant pathogens. The 'inducer' for this form of control may use a particular strain of the plant pathogen that has low virulence, a different species of microorganism or a natural product, as well as the plant itself. This is very different from the microbial control approach used against insects, which currently relies exclusively on using virulent parasites to directly kill insect pests.

Many microbial antagonists of plant pathogens have more than one way of restricting the development of a target pest. A number of species of the fungal control agent *Trichoderma*, for example, are used against soil-borne plant pathogenic fungi. *Trichoderma* species are able to parasitize plant pathogenic fungi in the soil, they also produce antibiotics and fungal cell-wall-degrading enzymes, they compete with soil-borne pathogens for carbon, nitrogen and other factors, and they can also promote plant growth, possibly by the production of auxin-like compounds (Verma *et al.*, 2007; Vinale *et al.*, 2008). *Trichoderma* is a common soil fungus and naturally grows in the rhizosphere. Multiple modes of action confer many benefits in terms of disease control, because *Trichoderma* gives good control in a range of conditions. However, it can create problems for the authorities that have to regulate its production and use. Many *Trichoderma* products have been sold on the basis of their plant growth-promoting properties, rather than as plant protection products, and so have escaped scrutiny from regulators in terms of their safety and efficacy.

At the microbial scale, plants present a very diverse set of environments for the microorganisms that are associated with them. The environmental conditions on the leaf surface are very different to those in the root zone, for example. The leaf surface is devoid of many microorganisms as conditions are not conducive to growth and survival. Water and nutrients are in scarce supply, while low humidity and high levels of ultraviolet radiation limit the germination of fungal and bacterial spores. In contrast, the root zone has freely available water and is bathed in large amounts of readily utilizable carbon secreted by root cells. As a result, there are large populations of taxonomically diverse microorganisms inhabiting the root zone and competing for resources. It is critical, therefore, that the ecology of the plant pathogen is understood in detail if biocontrol is to be successful.

Microbial antagonists
A number of microbial antagonists are being used as commercial products against plant pathogenic fungi and oomycetes (Table 3.1). Microbial control products have been developed for use against soil-borne plant pathogens and pathogens that infect the above-ground parts of plants. The most widely used fungal control agents in the soil are species of *Trichoderma*, such as *Trichoderma harzianum*, which is an antagonist of *Rhizoctonia*, *Pythium*, *Fusarium* and other soil-borne pathogens (Harman, 2005). *Trichoderma* is a parasite of a range of fungi and oomycetes in the soil, but it also inhibits the growth of other organisms by the production of toxic metabolites and cell-wall-degrading enzymes. Specific recognition reactions between parasite and host mediate the release of antimicrobial metabolites by the parasite. Other fungal parasites and antagonists include *Gliocladium virens* and *Coniothyrium minitans*. The latter is applied to the soil to kill *Sclerotinia sclerotiorum*, an important disease of many agricultural and horticultural crops such as oilseed rape, lettuce, carrots, beans and brassicas (Whipps *et al.*, 2008). Bacterial agents can also be used for control of soil-borne diseases. Crown gall, caused by *Agrobacterium tumefaciens*, is a serious disease of a wide range

of dicotyledonous plants including pome fruits, vines, ornamentals and vegetables. Bacterial infection causes the formation of tumours in root tissue. Seed and seedlings can be treated with the K84 strain of the non-pathogenic species *Agrobacterium radiobacter*. K84 colonizes root tissues and prevents occupation by *A. tumefaciens*, using an antibiotic (Penalver *et al.*, 1994). Specific strains of *Bacillus subtilis* can also confer protection against some root pathogens, while a number of *Pseudomonas* species, including *Pseudomonas fluorescens* and *Pseudomonas aureofaciens*, reduce damping off and soft rots (Kloepper *et al.*, 2004; Haas and Défago, 2005; Choudhary and Johri, 2009).

Fungal antagonists used against pathogens that infect leaves and stems include: *Lecanicillium*, which is primarily an insect pathogenic fungus, but some strains have activity also against other fungi; *Ampelomyces quisqualis*, which is used against mildews; and *Nectria inventa* and *Gonatobotrys simplex*, which are parasites of *Alternaria* (Kiss *et al.*, 2004). The fungus *Phlebiopsis gigantea* is used to control *Heterobasidion annosum*, a fungal pathogen that causes rots in freshly cut stumps of pine trees and which can spread subsequently to intact trees by root-to-root contact. *Phlebiopsis* spores are painted on to tree stumps or are incorporated in the lubricating oil used in chainsaws. The fungus occupies the same tissues as *Heterobasidion* and outcompetes it, but causes no damage to the trees (Pratt *et al.*, 1999).

There are also a number of bacterial species that are used as control agents of plant pathogens infecting above-ground parts of plants. Species of *Bacillus*, *Pseudomonas* and *Streptomyces* can prevent colonization of leaf and stem tissue by plant pathogens (Berg, 2009). The activity of these agents is often due to antibiosis brought about through the action of bacterial secondary metabolites. Usually, several kinds of secondary metabolites are produced. Production of metabolites is strain dependent, i.e. different strains of the same species of bacterium can produce different types of metabolites with different effects on target pathogens. Strain selection is therefore a critical part of developing bacterial agents as biopesticides.

Microbial antagonists can be formulated as dusts, granules or liquid suspensions for application to soil, either directly to the roots of plants or in the soil ahead of planting. Antagonists used on leaves, stems or harvested fruit are usually applied as conventional sprays. However, novel application systems are also being developed. Honeybees, used commercially for pollination of blueberries, transport the plant pathogenic fungus *Monilinia vaccinii-corymbosi* between blueberry flowers, leading to berry disease. However, the risk of the disease can be reduced significantly by using the bees as 'flying doctors' and treating them with the bacterial biopesticide *B. subtilis*, which is dispensed from a device fitted to the entrance of bee hives and which the bees vector to blueberry flowers (Dedej *et al.*, 2004).

Microbial antagonists are also used as control agents of postharvest diseases, mainly against the causal agents of rots in harvested fruits and vegetables. Yeasts, filamentous fungi and bacteria have all been used (Spadaro and Gullino, 2004). Mode of action is not always clear, although competition with the pathogen for space and nutrients is thought to be important, alongside antibiosis. Sharma *et al.* (2009) list over 40 species of microbial antagonists that have been demonstrated in experiments to give successful control of postharvest disease of fruits and vegetables. Common target pathogens in these experimental programmes included *Botrytis cinerea*, *Penicillium* species (e.g. *Penicillium digitatum*, *Penicillium expansum*, *Penicillium italicum*) and *Mucor piriformis*. At present, however, only nine products are available commercially across the world (Sharma *et al.*, 2009). Of these, the most widely used are based on *A. quisqualis* and *B. subtilis*. Application of the control agent may be made preharvest, to combat latent infections acquired in the field, although it is not considered a commercially viable strategy. Postharvest application is more practical, and the inoculum is usually applied as sprays or as a dip.

Developers of microbial biopesticides of plant pathogens have tended, quite understandably, to concentrate on species of microbial antagonists that are easy to culture and mass-produce, although it has been pointed out (Whipps and Davies, 2000; Alabouvette *et al.*, 2006) that if commercialization of these agents is done according to a chemical pesticide model, without proper consideration of the ecological interactions involving the control agent, the target pathogen, the crop plant and the environment, then poor or inconsistent levels of control are certain to occur, which would be damaging to the whole concept of biological control.

Microbial control of plant parasitic nematodes
Plant parasitic nematodes are susceptible to fungal and bacterial pathogens, a small number of which are available as commercial biopesticides. Nematophagous fungi include species that trap motile nematodes in the rhizosphere using specialized hyphal organs, such as *Arthrobotrys oligospora* and *Arthrobotrys dactyloides*. Endoparasitic fungi, such as *Pochonia chlamydosporia* (= *Verticillium chlamydosporium*) are able to infect female cyst nematodes and their eggs (Kerry, 2000). *Pochonia chlamydosporia* can be mass-produced *in vitro*, and some strains are able to grow saprotrophically within the rhizosphere, making it a potentially valuable augmentation biopesticide. However, the development of microbial control agents of plant parasitic nematodes appears to be relatively slow. Dong and Zhang (2006) list only nine products that have been commercialized, based on six different fungal or bacterial species. They attribute this lack of products partly to inconsistent performance in the field, and it is likely that product development is being held back by a lack of knowledge of the complex interactions that occur between nematode, control agent, the plant and the soil, and in particular the rhizosphere.

Using microbes to induce plant defence responses against pathogens
Plants have both constitutive and inducible forms of defence against plant herbivores and pathogens. When a plant is infected by a pathogen, it is often able to detect it and mount a defence by producing a range of antimicrobial compounds. Such induced responses are thought to have a selective advantage over constitutive defences because they reduce trade-offs between the allocation of resources for defence and other processes, such as growth and reproduction. Induced defences are initiated by elicitor molecules from the

pathogen, which include substances such as pathogen toxins, proteins and cell-wall constituents, such as glycopeptides, chitin and glucans. Binding of the elicitors to receptor molecules located on plant cell membranes initiates a biochemical cascade that results in the up-regulation of genes controlling the production of a range of defence chemicals, such as phytoalexins and antimicrobial proteins. Basal-level induced immunity to virulent pathogens is afforded by a non-specific pathogen recognition system that utilizes surface receptors (so-called pathogen-associated molecular pattern (PAMP)-triggered immunity, PTI) (Vlot et al., 2009). Plants also exhibit induced responses that are specific to particular pathogens, known as effector-triggered immunity (Vlot et al., 2009). These specific responses are rapid and often take the form of a hypersensitive reaction, in which reactive oxygen species and pathogenesis-related defence proteins are synthesized by the plant at the site of infection, while the plant also causes the death of infected cells in order to contain the pathogen and prevent the disease spreading. The hypersensitive response only occurs in specific host–pathogen combinations and requires that a particular host resistance protein recognizes a matching avirulence protein from the pathogen.

Two key discoveries about plant inducible defences have enabled them to be exploited for pest management (Walters et al., 2007): (i) inducible defences can be triggered in plants by avirulent strains of a pathogen, by non-pathogenic microbes or by chemicals; and (ii) inducible defences are not just confined to the site of contact with the elicitor, but are also induced in distal parts of the plant via a chemical message. This helps to protect the plant against subsequent infection by other plant pathogens. Two different types of plant inducible defence have been characterized based on differences in the chemical signalling pathways involved. They can both be induced artificially by treating the plant with microbial agents or with chemical elicitors. Once induced, long-lasting partial resistance is conferred against a wide range of diseases. Systemic acquired resistance is induced in nature by necrotrophic pathogens. The resistance first develops as a localized hypersensitive response and is induced elsewhere in the plant by chemical signals based on salicylic acid. Systemic acquired resistance can also be induced using specific chemical agents, some of which are available on a proprietary basis. In contrast, induced systemic resistance is triggered in plants by non-pathogenic, plant growth-promoting rhizobacteria. Induced systemic resistance is regulated by jasmonic acid and ethylene pathways. Induced systemic resistance acts independently of systemic acquired resistance although the two mechanisms share a common regulatory protein.

Both induced systemic resistance and systemic acquired resistance are exploited commercially. For example, the synthetic compound probenazole has been used in Asia since 1975 to manage the rice blast pathogen *Magnaporthe grisea* using systemic acquired resistance (Walters and Fountaine, 2009). Probenazole is applied during seedling production or to rice fields after planting, where it is absorbed by plant roots and translocated to the rest of the plant, and can control rice blast for up to 70 days (Iwata, 2001).

Other systemic acquired resistance products include acibenzolar-S-methyl (ASM) and chitosan, a polymer present in insect cuticle and fungal cell walls (Walters and Fountaine, 2009). In practice, the level of disease protection conferred by these products can vary widely, and is believed to be dependent upon a broad range of complex, interacting factors including plant genotype, environmental conditions, crop nutrition and whether the plant has already been induced by naturally occurring biotic and abiotic factors in the field prior to the application of the compound (Walters and Fountaine, 2009).

Symbiotic associations between arbuscular mycorrhizal fungi and plant roots are widespread in nature and improve plant nutrition, drought tolerance and pest resistance (Bending et al., 2008). Many agricultural crops are mycorrhizal and there is widespread if equivocal evidence that crops benefit from the association by induced resistance to plant diseases (Gosling et al., 2006). Intensively managed agro-ecosystems are generally impoverished in arbuscular mycorrhizal fungi. However, new approaches could be developed to reverse this, such as changes to farm management practices or applications of fungal inoculum to the soil. This could be a very practical way of boosting the ability of crop plants to withstand disease.

Microbial control of weeds

Plant pathogens can be exploited as microbial control agents of weeds. Biological weed control has been done successfully using the augmentation strategy (where the agents are referred to as 'bioherbicides') and classical biological control. However, only a small number of examples have been put into practice to date. Synthetic chemical herbicides are inexpensive, effective and widely available, which makes it difficult for bioherbicides to compete unless they can find a niche market. Between 1980 and 2001, only eight bioherbicides were registered worldwide (Charudattan, 2001). However, although the market is small, it has been argued that biological weed control must be considered alongside all other potential management tools given the diversity and extent of weed problems (Charudattan, 2001), and it has a particular role for key weed species that have evolved herbicide resistance, such as blackgrass *Alopecurus myosuropides* in Europe and annual ryegrass *Lolium rigidum* in Australia (Hallett, 2005). Part of the reason for the lack of commercial bioherbicide products may be the fact that there are few academic or commercial organizations in the world with the critical mass of scientific expertise necessary to go all the way from basic research to product development (Ash, 2010).

No bioherbicide products are currently available in Europe, but two products, Collego and DeVine, have been used in the USA since the 1980s. These are based on strains of the fungus *Colletotrichum gloeosporioides* and the oomycete *Phytophthora palmivora* (Charudattan, 1990). Collego has been sold since 1982 as a bioherbicide of northern jointvetch, *Aeschynomene virginica*, in soybeans and rice (te Beest *et al.*, 1992). Northern jointvetch causes seed contamination at harvest; it reduces rice yields by up to 20% and causes around 10% of the crop to be rejected because of low quality. The strain of *C. gloeosporioides* used in Collego has high virulence, is able to kill seedlings within several weeks and the commercial product gives over 90% control when used correctly by growers. It is integrated into rice IPM systems, which includes proper timing of fungicide applications to ensure that they do not inhibit the bioherbicide. *Phytophthora palmivora* has been used for nearly 30 years to control stranglervine (also known as milkweed vine), *Morrenia odorata*, in Florida citrus groves. Stranglervine is an alien, invasive species that was introduced to Florida from South America as an ornamental plant. It climbs up citrus trees and competes with them for water, light and nutrients. DeVine uses a highly virulent strain of *P. palmivora* isolated from diseased stranglervines in Florida (te Beest *et al.*, 1992). It is not specific to stranglervine, and in glasshouse experiments it was found also to infect a range of crop plants including pea, squash, watermelon, tomato and potato when applied at high inoculum levels in sterile soil. However, it was argued that these represented an extreme case and that in the field these plant species would not be exposed to direct applications at the same concentrations of inoculum. DeVine is highly effective; an application provides 95–100% control that persists for at least 1 year (Kenney, 1986; te Beest *et al.*, 1992).

One of the first notable examples of using a pathogen for classical weed control concerns the rust fungus *Puccinia chondrillina* for control of skeleton weed *Chondrilla juncea*, a serious invasive weed of agriculture in south-east Australia (te Beest *et al.*, 1992). There is a very high degree of specificity between the weed host and the pathogen, so that each strain of the weed is susceptible only to a particular co-evolved strain of fungus. An Italian strain of the fungus that was highly virulent to the weed was released in 1971, and it spread rapidly after release. Two Italian strains were also released in California and Oregon in the USA for control of skeleton weed there. These pathogen strains have now colonized the entire Pacific north-west. Pathogen infection reduces plant biomass and seed production, and 90% weed mortality can occur when conditions are favourable (te Beest *et al.*, 1992). Another rust fungus, *Uromycladium tepperianaum*, was introduced to control the Australian acacia tree, *Acacia saligna*, a serious invasive weed in the Cape Fynbos in South Africa, a region with a unique and particularly diverse native flora. The fungus was introduced in the late 1980s and reduced tree density by 95% in 9 years (Charudattan, 2001). Classical control of weeds has also proved beneficial in Hawaii, where five major biological control projects against weeds have been done since the late 1960s (Trujillo, 2005). The economic and ecological benefits of these programmes have been considerable. For example, control of just one weed species, *Senna surattensis*, with a bioherbicide based on the fungus *Acremonium*, is reported to have saved the Hawaiian cattle industry from collapse in the 1970s, and the total economic benefit of the Hawaiian weed biocontrol programme is valued at greater than US$100 million (Trujillo, 2005).

Natural Compounds

Animals, plants and microorganisms synthesize a wide array of chemical compounds that have adaptive significance in terms of defence against predators, locating sources of food, finding mates, etc. The importance of some of these compounds to the ecology of plants, pests and their natural enemies is now relatively well characterized, and methods for exploiting them within IPM are being developed.

Botanicals: direct chemical defences used by plants against herbivores

Plants produce a wide variety of secondary metabolites that deter herbivores from feeding on them. These chemicals include terpenoids (for example volatile oils such as limonene), phenolic compounds (such as tannins), alkaloids (such as nicotine) and glucosinolates (such as mustard oil). When extracted and concentrated, some of these plant compounds can be used as pesticides. About 50 different botanical active substances are registered as plant protection products in the USA, but only 11 are registered in the EU (Table 3.1).

We have mentioned previously that one of the first commercial pesticides to be introduced was nicotine. Nicotine is a potent non-specific insecticide that is synthesized by plants from the genus *Nicotiana* (including the commercial tobacco plant *Nicotiana tabacum*) as a defence against insect herbivores. It binds to acetylcholine receptors and affects neurotransmission. At normal concentrations in *Nicotiana* plants it acts as a feeding deterrent, but is insecticidal at the higher concentrations found in *Nicotiana* plant extract. Nicotine is also a potent mammalian toxin, which goes to illustrate that just because a pesticide is natural, it does not necessarily mean that it is safe. However, there are other phytochemical insecticides which are safe to use and which are exploited commercially. They are often referred to as 'botanicals'. The market for botanical pesticides is currently limited, probably because of the low efficacy of the products and the expense of mass production compared with synthetic chemical pesticides. In developed countries, they tend to be used most in organic farming. The most widely used botanical compound is neem oil, extracted from seeds of the neem tree, *Azadirachta indica*, which is native to the Indian subcontinent (Schmutterer, 1990). Neem oil contains two components with insecticidal activity, azadirachtin and salannin. They function both as an insect feeding deterrent and a growth regulator, preventing moulting in immature insects that ingest them (Atawodi and Atawodi, 2009). Neem oil has been used as a traditional insecticide in India for over 4000 years. Neem oil has a low mammalian toxicity and has some systemic activity when applied to plants. A number of neem oil insecticidal products are used in India and elsewhere. However, neem oil is expensive to mass-produce compared with synthetic chemical pesticides, and it has a low efficacy. Koul (2008) estimates that 10–30 kg of neem seeds are needed to treat 1 ha of land, at a cost of US$5–20/ha. Elsewhere, research on the effectiveness of low-molecular-weight volatile plant oils, also known as essential oils, against insect and mite pests has been done with oils from over ten different plant species. For example, essential oil from *Chenopodium ambrosioides* was shown to be effective at controlling fungus gnats, *Bradysia coprophila*, on ornamental plants (Cloyd and Chiasson, 2007). The efficacy of commercially available products against arthropod pests can vary depending upon the active ingredient, and while some products can give high levels of control, others may have low efficacy or be phytotoxic (Cloyd et al., 2009). It is important therefore that independent evaluations are made of the effectiveness of these products including phytotoxicity.

Other botanical compounds included the pyrethrins, which are the active compounds of pyrethrum, a powder produced from flowers of *Chrysanthemum cinerariaefolium*. Pyrethrins kill most insect species, are active at low concentrations and give a rapid knockdown effect (Silverio et al., 2009). The product has a low mammalian toxicity, but as we have mentioned previously it is toxic to aquatic animals and it also degrades rapidly after application. While this is an attractive property from an environmental point of view, as it reduces the chances of non-target organisms being affected, it does lower the overall effectiveness against target insects. The poor persistence of pyrethrum was the main driver for the development of synthetic forms of the chemical with better environmental performance. Rotenone is a flavonoid insecticide active against leaf-feeding insects and extracted from tropical legumes in the genera *Derris* and *Lonchocarpus* (Guleria and Tiku, 2009). Rotenone works as a contact and stomach-acting insecticide. However, it is moderately toxic to mammals and very toxic to fish.

Pesticides based on microbial products

Two highly active pesticides are available based on microbial products. Strictly speaking they fall within our definition of a biopesticide; however, the regulatory authorities have evaluated and approved them as conventional chemical pesticides. Spinosad is a mixture of two macrolide compounds, spinosyn A and D, that are synthesized by a soil actinomycete, *Saccharopolyspora spinosa*, which was collected from a rum still in the Caribbean (Mertz and Yao, 1990). The level of insecticidal activity is comparable to that of synthetic pyrethrins, it has a very low mammalian toxicity and residues degrade rapidly in the field. A review of studies of the impact of Spinosad on natural enemies showed that it is not harmful to predators but that it can be moderately harmful to parasitoids (Williams et al., 2003). Spinosad kills insects by causing hyperexcitation of the central nervous system (Sparks et al., 2001). Farmers and growers used it widely against western flower thrips, lepidopteran and dipteran pests following its introduction in 1997, but unfortunately resistance has already developed in some of these pests (Herron and James, 2005; Bielza et al., 2007). In the diamondback moth, one of the most important pests of brassicas worldwide, resistance is determined by an incompletely recessive gene mutation (Baxter et al., 2010).

Abamectin belongs to a class of macrocyclic lactone compounds, the avermectins, that are synthesized by the soil actinomycete *Streptomyces avermitilis* strain MA-4860 isolated from Japan in 1976 (Lasota and Dybas, 1991). These compounds affect sodium-gated membrane channels in nerve cells that utilize γ-aminobutyric acid as a neurotransmitter. Abamectin is degraded rapidly in the environment, including in the soil, in water and on plant surfaces, it is not bioaccumulated in the food chain and therefore is considered not to represent an environmental hazard. It is active against a range of pest species but it has particularly good activity against mites, including many species, such as the two-spotted spider mite and the citrus rust mite, which could not be controlled with conventional acaricides because of resistance. Unfortunately, resistance has developed now also to abamectin (Sato *et al.*, 2005; He *et al.*, 2009).

Semiochemicals

A semiochemical is a chemical signal produced by an organism that evokes a change in the behaviour of an individual of the same or a different species. Semiochemicals are divided into two broad classes, allelochemicals and pheromones. Allelochemicals are substances produced by one species and which cause a response in individuals of another species. They are categorized according to the effects they have on the emitter and the receiver: (i) allomones are allelochemicals that are beneficial to the organism that has emitted them, but are detrimental to the receiving organism; (ii) in contrast, kairomones are beneficial to the receiver and detrimental to the emitter; while (iii) synomones are attractive to both emitter and receiver. Pheromones are chemicals released by an organism that affect individuals of the same species. Both pheromones and allelochemicals are being used for pest management.

Exploitation of allelochemicals against invertebrate pests
Plants produce volatile allelochemicals in response to feeding or egg laying by insect herbivores (for this reason they are often referred to as herbivore-induced plant volatiles, HIPVs). They include methyl jasmonate, methyl salicylate and hexenyl acetate. Note that both the jasmonate and salicylate compounds were discussed earlier in this chapter in relation to plant defence against pathogens. These signalling molecules are produced by a taxonomically diverse range of plant species, which suggests that they constitute an evolutionarily ancient form of plant defence. In the same way that they act as a messenger to prime distal parts of the plant against pathogen infection, they also spread systemically from wound sites induced by invertebrate feeding, stimulating the synthesis of phytochemicals that deter feeding and oviposition in other parts of the plant. They also stimulate the production of deterrent compounds in neighbouring plants (Thaler, 1999; Reddy and Guerrero, 2004). They are also able to recruit natural enemies as 'bodyguards' when under attack: insect predators and parasitoids have evolved the ability to detect HIPVs and use them to detect prey. Different plant species produce

different blends of volatiles according to the type of herbivore that feeds upon them.

This knowledge can be exploited for pest control. For example, James *et al.* (2005) showed that stick traps baited with hexenyl acetate, methyl jasmonate or methyl salicylate increased the abundance of natural enemies on the crop. The precise mechanism by which natural enemies are attracted has yet to be elucidated, but it is hypothesized that crop plants respond to a general warning signal from the HIPV and emit their own specific blends of volatiles to attract in natural enemies.

Insect pheromones

There are three main functional types of insect pheromone: (i) sex pheromones, produced by one sex to attract the other; (ii) aggregation pheromones, which cause individuals to group together for the purposes of reproduction, feeding or hibernation; and (iii) alarm pheromones, such as the sesquiterpene β-farnesene, which is produced by aphids when attacked by predators or parasitoids and causes the dispersal of other aphids in the vicinity (Kunert *et al.*, 2005). Nowadays, it is possible to synthesize a wide range of insect pheromones and use them for crop protection. These synthetic molecules are chemically identical to the ones found in nature. Each insect species has its own particular blend of sex pheromones; so considerable research effort may be required to produce exactly the right blend of synthetic pheromones for use in the field. Sex pheromones are used for both monitoring and pest control. For monitoring, a small number of lures are deployed in the field, each comprising a pheromone dispenser with a sticky trap. The numbers of trapped insects provide information on flight activity, dispersal, population size and so on. The data are used to inform the timing of applications of pesticide sprays or other therapeutic agents. Sex pheromones are also used for pest control. This can take the form of mass trapping, where pheromone dispensers attract insects to physical traps. Mass trapping is used, for example, to control banana weevil in commercial banana plantations (Reddy *et al*, 2009). Sex pheromones can also be utilized in lure-and-kill systems, in which the pheromone is used to attract the insect to a device containing a pesticide (El-Sayed *et al.*, 2009). The advantage of this system is that the amount of pesticide used is significantly less than in conventional spraying, and keeping the pesticide within the lure means that there are no issues of contamination of the crop plant. There are also lure-and-infect systems, in which the pest is treated with a microbial pathogen such as a baculovirus or a fungus. However, the most highly developed and widely used strategy for using sex pheromones is mating disruption. Large numbers of female sex pheromone dispensers are deployed within a field to disorientate males and disrupt their ability to find females for mating. Worldwide, mating disruption is used on over 660,000 ha (Witzgall *et al.*, 2008). One of the most successful examples is the use in apple orchards of the codling moth female sex pheromone, (E,E)-8,10-dodecadienol (known as codlemone), which is used on about 160,000 ha worldwide. Because there is a low threshold for economic damage to the apple crop, the pheromone is most effective at low to medium population levels. If the moth population exceeds 1000 overwintering larvae per hectare then the pheromone needs to be combined with other management tools, such as the codling moth baculovirus biopesticide (Witzgall et al., 2008). Insect pheromone formulations are non-toxic to mammals and are recognized as low-risk products by regulatory authorities. The US EPA, for example, does not require toxicological testing for straight-chain lepidopteran pheromones (SCLPs) if applied at less that 150 g/acre (Weatherston and Stewart, 2002). Not all insects produce pheromones that can be utilized for pest management, and even if a pheromone is available there may only be limited opportunities to control a pest due to agronomic factors or the pest's behavioural ecology. However, sex pheromones clearly have much to offer in terms of making crop protection more sustainable.

Use of Biopesticides in Integrated Pest Management

Modern IPM is an ecologically based discipline. Successful IPM programmes should take into account the following: the biological characteristics of the pest; the causes of the pest outbreak; the presence of other pests; the attributes of the available control agents including their safety, efficacy, environmental impact and compatibility with other IPM tools; the characteristics of the crop and its environment; the financial costs of the available pest control tactics; and social factors (for example, are the available control strategies acceptable to society?). It must be remembered that the number and diversity of species that make up the pest community on a particular crop are critically important in determining the overall pest management strategy. Writing about the use of microbial biopesticides in IPM, Lacey and Shapiro-Ilan (2008) point out that pest management agents tend to be considered as stand-alone treatments and that the ways in which they interact with the rest of the agroecosystem are often overlooked. This would appear to be a hangover from the industrial approach to crop protection. Some biopesticides work well as stand-alone treatments, such as *M. anisopliae* used for control of locusts and grasshoppers (Lomer et al., 2001). But many biopesticides, by their nature, are unlikely to work effectively as stand-alone treatments; however, their selectivity and safety mean that they can contribute meaningfully to incremental improvements in pest control in ways that make agriculture more sustainable (Lacey and Shapiro-Ilan, 2008). The development of systems that integrate biopesticides with other control measures is essential for the future of IPM. This vital work is complex, challenging and long term, but it is also true to say that it is unglamorous and does not receive the attention it deserves from governments and other research funders.

We have already seen that sophisticated IPM systems are used widely in protected crops, particularly for the control of insect and mite pests. In Europe, the industry is supported by about 25 biological control companies – including the world's three largest – that supply natural enemies and technical support (van Lenteren, 2000). The costs of IPM are reported to be competitive with chemically based control (van Lenteren, 2000). Fungal and nematode biopesticides are incorporated into these systems. As an illustration, a leading grower of glasshouse ornamental crops in the UK routinely uses 11 different biocontrol agents including parasitoids (*Aphidius, Dacnusa, Diglyphus, Encarsia*) and predators (*Aphidoletes, Amblyseius, Hypoaspis, Phytoseiulus*) as well as entomopathogenic nematodes and the insect pathogenic fungus *Lecanicillium*. The system is based on routine, detailed monitoring of pest abundance, and a computerized system is employed to keep a check on the amount and cost of biocontrol used per week, month and year for each crop grown. Where possible, mechanized systems of biocontrol application are used to save costs. The whole system keeps application of chemical pesticides to a minimum and works out at about three pence per plant (M. Holmes, personal communication).

Research has shown that biopesticides can be used as an effective second line of defence with predators against western flower thrips and spider mites on greenhouse crops (Jacobson *et al.*, 2001; Chandler *et al.*, 2005). These pests are routinely managed using trickle release applications of predatory mites. However, there are invariably periods in the production season when the pest population starts to outstrip the ability of the predator to control it. This may be caused by a change in environmental conditions such as temperature, the application of a chemical pesticide, or because some crops such as tomato (which has leaves covered in thick, glandular trichomes) are not very conducive to easy predator movement. In the past, synthetic chemical pesticides have been used in such situations, but their use has become severely restricted through the evolution of resistance. However, this gap can be replaced by using biopesticides in some cases. For example, the entomopathogenic fungus *B. bassiana* applied as a supplement to the predatory mite *Phytoseiulus persimilis* against spider mites, *T. urticae*, can give up to 97% control of eggs, nymphs and adults seven days after spraying (Chandler *et al.*, 2005). Growers in countries where *Beauveria* is available commercially are now adopting this strategy. The ability to use fast-acting, supplementary treatments such as *Beauveria* can make the difference between success and failure in greenhouse IPM (R. Jacobson, personal communication).

Perhaps the key point to make about IPM in greenhouse crops is that multiple management tactics, including biopesticides, are integrated across a wide range of different groups of pests and are also integrated with agronomic practice. This makes for a highly advanced system. For other crop sectors, IPM – where it is practised at all – is still at the stage of integrating different pest control tactics for the same pest or group of pests. Getting biopesticides to work effectively for crops grown outdoors is significantly more of a challenge than for crops grown under protection. An important area for future activity is going to be making biopesticide products available for field crops to be used in IPM. In Chapter 2 we saw how Bt has been incorporated into a biocontrol-based IPM system for outdoor tomato production in the USA. Bt can also be a valuable tool for growers against caterpillar pests on a range of other field vegetable crops including brassicas, sweetcorn, pepper, aubergine and potato (Rowell and Bessin, 2005). The fact that the Bt products used have no negative effect on the predator and parasitoid natural enemies of caterpillar pests is an important plus point for IPM. The mycoparasitic fungus *C. minitans* is used as a biopesticide of sclerotinia disease caused by *S. sclerotiorum*, which affects a wide range of crops. At present the mycoparasite is used mainly in greenhouses, but the future aim of the various biopesticide companies involved with it is to market it also for outdoor use including on oilseed rape, for which sclerotinia can be a major problem. Integrating *Coniothyrium* into IPM will be essential. Its effectiveness can be improved in IPM, for example by combining its application with organic amendments such as lignin or with reduced fungicide applications (van Beneden *et al.*, 2010).

There are encouraging prospects for the incorporation of biopesticides into IPM systems in orchard and soft fruit crops. Because these crops are eaten raw and unprocessed, there is considerable demand from retailers and consumers for reduced use of chemical pesticides. On the supply side, a range of reliable products is becoming increasingly available. Bt, baculoviruses and entomopathogenic nematodes are being used against a range of lepidopteran pests of pome and stone fruits such as tortrix moths and fruit borers. The entomopathogenic nematodes *Steinernema carpocapsae*, *Steinernema riobrave* and *Heterorhabditis indica* are being used for management of citrus root weevils (Coleoptera) in the USA. Granulosis virus-based biopesticides have been available since the 1980s against codling moth, *C. pomonella* (Lepidoptera), which is the most important pest of apples worldwide (Lacey and Shapiro-Ilan, 2008). Codling moth larvae burrow into developing apples to feed, causing significant economic damage to the crop if left uncontrolled. The CpGV virus is applied against neonate larvae before they have entered the fruit. It is highly virulent and has a rapid speed of kill, although some cosmetic damage may still occur from larvae that burrow into the fruit surface before death. Commercial CpGV products are available in most European countries, in North and South America, as well as in South Africa, Australia and New Zealand (Jehle, 2008). Resistance management is required as some codling moth populations have been reported in Europe following repeated applications over many years (Asser-Kaiser *et al.*, 2007). Integrated systems are being used successfully against resistant moths using a combination of CpGV with pheromones for mating disruption. This can be reinforced by the application of entomopathogenic nematodes to soil, leaf litter, etc. to control overwintering diapausing pupae (Lacey and Shapiro-Ilan, 2008).

Biopesticides are not being used much at all on broad-acre crops in industrialized countries, as in general they cannot yet compete with cheaper, more effective synthetic pesticides. Pest management in broad-acre crops in these countries is based largely on stand-alone tactics for each pest, and hence the lack of a pre-existing IPM framework is likely to be a significant barrier to biopesticide uptake. However, where pesticide use becomes untenable, then biologically based IPM is adopted. As an example, alternative control mechanisms play an important role in Colorado potato beetle management in North America. The most widely used natural enemies are the entomopathogenic fungus *B. bassiana* and sprays of Bt, and there is evidence that the two entomopathogens work synergistically when applied together (Wraight and Ramos, 2005). The egg parasitoid *Edovum puttleri*, used according to an augmentation strategy, can give about 50% parasitism in the field. Because overwintering adults disperse to the crop by walking, and hence do not cover large distances, crop rotation can be a viable form of cultural control (Hare, 1990). Despite considerable work on breeding for resistance, no conventionally bred resistant crops are available. Crops engineered to express Bt endotoxin were made commercially available in the USA in 1995 but were withdrawn from use in 2000 because of consumer concerns about GM crops (Romeis et al., 2006). Research is under way to characterize wild relatives of the cultivated potato that have Colorado potato beetle resistance and which could be used in a crop breeding programme. One promising finding has been that resistance to Colorado potato beetle is associated with particular taxonomic groups of *Solanum*, which should make it more cost effective in future to screen the large number of candidate plant accessions available (Jansky et al., 2009). Resistant species tend to produce foliage with high levels of glycoalkaloids and a high density of glandular trichomes. However, there remains a significant challenge of how to achieve introgression of resistance genes from wild relatives into cultivated potato lines. Since high levels of glycoalkaloids can cause bitterness in potato tubers, it has also been suggested that Bt genetic modification technology is combined with conventional breeding. This may reduce the selection pressure for resistance to Bt genes and allow the deployment of breeding lines with levels of glycoalkaloids in the foliage that do not affect the palatability of potato tubers (Cooper et al., 2004).

Probably the best-known example of implementing an integrated approach to biological control on a large scale concerns Cuba. A detailed examination of the Cuban agricultural crisis including the development of the national IPM system has been written by Nicholls et al. (2002). In the early 1990s, a sudden collapse of trade with the former Soviet countries (which were supplying Cuba with chemical pesticides, fertilizers and petroleum), plus problems with pesticide resistance, resurgence and secondary pests, forced the abandonment of chemical-based agriculture and a large-scale shift to semi-organic production in order to prevent widespread starvation. To give some scale to the Cuban agricultural crisis, in just 3 years from 1989, petrol imports fell by 53%, fertilizer imports by 77% and pesticide imports by 63%. During the same period, the per capita food intake fell from 12.17 to 7.79 MJ/day (2908 to 1863 kcal/day); however, by 2000 this had been brought back up to 10.82 MJ/day (2585 kcal/day) – the recommended daily intake is 8.37 MJ (2000 kcal) for women and 10.46 MJ (2500 kcal) for men. Reduced access to oil for road transport as a result of the US trade embargo resulted in significant amounts of food being grown in urban areas. Urban farmers are trained in IPM and thus biologically based IPM is used both in cities and in rural areas. The new agricultural production system that was developed is characterized by low inputs of chemical pesticides, the use of resistant varieties, growing crops that are best suited to local environmental conditions, use of crop rotations and mixed cropping, and integrated crop and livestock systems. Habitat manipulations are put in place to make the environment less favourable to pest populations and more favourable to natural enemies. Natural enemies, including microbial biopesticides, are used very widely. These are produced locally using a network of state-run mass-rearing facilities. There are around 280 biocontrol agent production centres, most of which provide parasitoids, predators and microbial agents at low cost to local farmers. In addition there are one industrial and 29 semi-industrial scale factories providing commercial quality products for use on export crops. Biological control is based on inundative applications of local strains of biocontrol agents, including the release of 10,000 million *Trichogramma* parasitoids per year. Production of microbial agents – including Bt, entomopathogenic nematodes and five different genera of entomopathogenic fungi – amounts to about 2000 t/year applied to 600,000 ha. In general, the Cubans use low-dose applications of microbial biopesticides, which by itself gives only partial pest control but is considered to be cost effective and efficacious when combined with other IPM components. Bt is sprayed against lepidopteran pests, such as diamondback moth, cassava hornworm and tobacco budworm. *Beauveria bassiana* is used against banana root borer, sweet potato weevil, sugarcane borer and rice weevil. *Metarhizium anisopliae* is targeted against banana root borer, diamondback moth and wax moth, while *Lecanicillium* species are sprayed against whiteflies and aphids. The IPM system is carefully matched to the particular characteristics of each type of crop. Thus IPM in sugarcane is based around use of resistant varieties and releases of predators and parasitoids, whereas banana IPM makes high use of cultural pest control and microbial biopesticides. While it has been successful in preventing nationwide starvation, the Cuban agricultural system clearly does not provide the range and abundance of food that consumers are used to in industrialized countries, and the high levels of farm labour that are required are probably untenable in modern market-driven economies. However, it must not be dismissed out of hand. The key elements of the Cuban system – biocontrol with local strains of natural enemies, the use of resistant plant varieties, careful attention paid to crop nutrition and rotation, plus state support including training for growers – is a prime example of sustainable IPM that is not reliant on high inputs of fossil-fuel-based pesticides and fertilizers.

As we have alluded to earlier in this book, there are complex interactions between insects, plants and phytochemicals. For example, some groups of herbivorous insects have evolved the ability to utilize phytochemicals for their own benefit. Some species of Lepidoptera, Coleoptera and Orthoptera sequester toxic pyrrolizidine alkaloids (PAs) from their host plant. Enzymes within the insect haemocoel oxidize the alkaloid to a non-toxic state. PA sequestered during the larval stage remains within the insect after metamorphosis into the adult stage and is even passed on to eggs. The sequestered compound acts as a deterrent to invertebrate and vertebrate predators; when ingested, it is reduced in the gut and absorbed in its toxic state (Reddy and Guerrero, 2004). Some insect species also exploit phytochemicals to ensure that eggs are laid on suitable host plants. Volatile phytochemicals acquired when an adult contacts a host plant can trigger the production of sex pheromone to attract mates (e.g. in *Helicoverpa* spp., cotton boll weevil, *Anthomonus grandis*, and African palm weevil, *Rynchophorus phoenicus*) (Raina *et al.*, 1992). Some insect species even use phytochemicals as sex pheromones or pheromone precursors (Nishida, 2002). In other cases, there is a synergistic interaction between a pheromone emitted by an insect herbivore and the volatile compounds produced by its host plant.

This knowledge can be exploited for improved, integrated pest control. For example, chemical baits used against African palm weevil are made of a mixture of host plant esters and insect sex pheromone (Reddy and Guerrero, 2004). The complex interactions between plants, herbivores, natural enemies and semiochemicals are starting to be exploited in the so-called 'push–pull' strategy (Cook *et al.*, 2007; Khan *et al.*, 2008) which uses behavioural modification to make a crop unattractive to pests (the 'push') while simultaneously luring it away from the field to other, more attractive plants that are grown as traps/baits (the 'pull'). Once the pests are on the trap crop they can be removed. The methods used for the 'push' part of the strategy are based on making the crop plants repellent or masking their apparency to pests. They can include use of non-host-plant volatiles, anti-aggregation pheromones, alarm pheromones, feeding and ovipositional deterring chemicals, and distorting the visual ability of pests to locate host plants using intercropping or undersowing with non-host plants. The trap crop can be treated with a range of semiochemicals such as host plant volatiles, pheromones, ovipositional and feeding stimulants, etc. Each individual component may have only a small effect on pest behaviour, but the overall aim is for the combined effect to be as efficacious as using a broad-spectrum chemical pesticide. The strategy is appealing from an IPM perspective, since it minimizes the use of synthetic chemical pesticides and conserves natural enemies. The push–pull approach is currently being used by African farmers to control stem-boring caterpillars in maize and sorghum. Stands of sorghum or maize are intercropped with molasses grass, which produces volatile compounds that repel ovipositing adults and also increases the presence of insect parasitoids. Plantings of trap crops consist of Napier grass or Sudan grass. Napier grass appears to be particularly suitable as not only is it attractive to ovipositing females, when its tissues are penetrated by boring larvae it also secretes a gum that coats the larvae and kills them. To date, however, there is little evidence that the push–pull strategy is being used consistently in commercial crops elsewhere. This may be because the system requires a high level of detailed knowledge in order to develop it, can be disrupted by changing the crop cultivar grown, or because the high level of species specificity is a cost-limiting factor.

Regulating Biopesticides

Many governments have regulations in place for biopesticides. Only authorized biopesticide products can be used legally for crop protection. Biopesticide companies have to apply for state authorization of a product; they are then required to submit detailed information in the form of a dossier, which is assessed by the government-appointed regulatory authority. The data requirements used now by most OECD countries for microbial biopesticides, pheromones and other semiochemicals are summarized in Tables 3.3 and 3.4. Usually these data requirements have been modified from those used for the registration of conventional synthetic chemical pesticides. Until about the mid-2000s, government regulators – particularly in Europe – were unfamiliar

Table 3.3. Types of registration information required for microbial pest control agents (MPCAs) in OECD countries. (From OECD, 2003.)

Information, test or study	R or CR	EU	USA	Canada	Japan	Australia
Point 1: Identity of the MPCA						
Applicant	R	✓	✓	✓	✓	✓
Producer	R	✓	✓	✓	✓	✓
Scientific name	R	✓	✓	✓	✓	✓
Composition of technical grade of MPCA/active substance						✓
Concentration of microorganism	R	✓	✓	✓	✓	✓
Composition of microbial material used for manufacture of end use products. Identity and maximum content of microbial impurities	R	✓	✓	✓	✓	✓
Quality criteria for the production and storage of the MPCA	R	✓	✓	✓	✓	✓
Quality control data	R	✓	✓	✓	✓	✓
Theoretical discussion regarding impurities. Physical and chemical properties if produced as manufacturing product. International regulatory status. Comprehensive data summary. Sample of MPCA, analytical standard of metabolite	R	✓	✓	✓	✓	✓
Patent status	R			✓		
Point 2: Biological properties of the MPCA						
Origin, isolation, maintenance and history of the isolate	R	✓	✓	✓	✓	✓
Natural occurrence of microorganism	R	✓	✓	✓	✓	✓
Description of target organisms	R	✓	✓	✓		✓
Available information on host specificity	R	✓	✓	✓	✓	✓
Life cycle of microorganism	R	✓	✓	✓	✓	✓
Information on closely related species. Physiological properties. Description of plasmids. Genetic stability. Resistance to antibiotics	R	✓	✓	✓		✓
Point 3: Information on the MPCA (function, mode of action, handling)						
Function	R	✓		✓	✓	✓
Fields of use	R	✓		✓	✓	✓

Continued

Table 3.3. *Continued*

Information, test or study	R or CR	EU	USA	Canada	Japan	Australia
Details of existing and intended uses	R	✓	✓	✓		✓
Details of harmful organisms against which protection is afforded	R	✓	✓	✓	✓	✓
Effects achieved	R	✓		✓	✓	✓
Mode of action in terms of biochemical and physiological mechanism and biochemical pathways	R	✓		✓	✓	✓
Details of active metabolites and degradation products	R	✓	✓	✓	✓	✓
Information on formation of active metabolites and degradation products	R	✓	✓	✓		✓
Information on possible resistance developing	R	✓				✓
Material safety data sheet for microbial active substance	R	✓		✓		
Detailed instructions for safe disposal	R	✓	✓	✓		
Decontamination of water procedures in case of an accident	R	✓		✓		
Other/special studies	CR			✓		✓
Crops or products to be protected or treated	R	✓	✓	✓	✓	
Measures to render microorganism harmless, in case of an accident	R	✓				
Point 4: Analytical methods						
Methods to preserve and maintain master seed stock	R	✓	✓	✓	✓	✓
Production process for technical grade of MPCA	R	✓	✓	✓	✓	✓
Quality control and post-registration monitoring methods	R	✓	✓	✓		✓
Storage stability test, data and determination of shelf-life	R	✓	✓	✓	✓	✓
Post-registration monitoring methods to determine and quantify residues and metabolites on food, feed, animal tissue, soil, water and air	CR	✓				
Point 5: Toxicological and exposure data and information on the MPCA						
Summary of potential of hazards to humans	R	✓		✓		✓
Occupational health surveillance report on workers during production and testing of MPCA	R	✓	✓	✓	✓	✓
Acute oral infectivity and toxicity	R	✓	✓	✓	✓	✓

Continued

Table 3.3. Continued

Information, test or study	R or CR	EU	USA	Canada	Japan	Australia
Acute intratracheal/inhalation infectivity and toxicity	R	✓	✓	✓	✓	✓
Acute intravenous/intraperitoneal infectivity	R	✓	✓	✓	✓	✓
Cell culture study for viruses and viroids or specific bacteria or protozoa	R	✓	✓	✓	✓	✓
Genotoxic potential, especially for fungi and actinomycetes	R	✓		✓		✓
Toxicity studies on metabolites	CR	✓	✓		✓	✓
Published reports on adverse effects. Short-term toxicity. First aid measures	R	✓	✓	✓		✓
Other/special studies	CR	✓		✓	✓	✓
Summary of mammalian toxicity and overall evaluation	R	✓			✓	✓
Point 6: Metabolism and residue studies on the MPCA						
Summary of residue behaviour and rationale for waiver of residue data	R	✓	✓	✓	✓	✓
Point 7: Fate and behaviour studies on the MPCA in the environment						
Information on origin, properties, survival and residual metabolites to assess fate in environment	EU – R USA – CR Japan – CR Canada – CR	✓	✓	✓		
Other/special studies	CR			✓		
Point 8: Ecotoxicological studies on the MPCA						
Effects on non-targets	R	✓	✓		✓	
Birds	R	✓				
Fish	R	✓	✓	✓	✓	
Aquatic invertebrates	R	✓	✓	✓	✓	
Effects on algal growth	R – EU only	✓				
Effects on aquatic or terrestrial plants	CR	✓	✓	✓		✓
Bees	R	✓	✓	✓		✓
Non-target terrestrial arthropods	R	✓	✓	✓		✓
Other terrestrial invertebrates	CR	✓		✓		
In EU effects on earthworms required	R	✓				
In EU effects on non-target soil microorganisms required	R	✓			✓	
Other/special studies	CR	✓				
Point 9: Summary information for the MPCA						
Summary and evaluation of environmental impact and assess risk	R	✓		✓		

R, the data submission is in principle necessary; CR, the data submission is necessary when the microbial pesticide meets a certain criterion.

Table 3.4. Guidance for registration requirements for pheromones and other semiochemicals used for arthropod pest control. (From OECD, 2001.)

Information, test or study	R or CR
Mode of action	
Function, handling and label information	
Information on function, directions of use, formulations, field of use and use sites, pests controlled, application rate, method and timing, preharvest interval, precautionary and emergency measures, procedures to clean equipment and spills, disposal of unused product	R
Labelling requirements regarding hazard classification and risk identification	R
Chemistry	
Technical grade of active ingredient (TGAI)	
Composition	R
Identity by spectral confirmation	R
Description of starting materials, production process and potential impurities	R
Analytical data and methodology	R
Analytical methodology and data for impurities of toxicological concern	CR
Analytical methods for residues	CR
Colour, odour, physical state, relative density or specific gravity, stability	R
For each known active ingredient (AI) component of the TGAI	
Description of starting materials and manufacturing process	R
Physical properties: melting point, boiling point, solubility in water and other solvents, colour, odour	R
Ultraviolet/visible absorption	R
Vapour pressure	R
Volatility (Henry's law constant)	R
Dissociation constants	R
Octanol/water partition coefficient	R
Submission of analytical standards (samples)	R
End-use product (EP)	
Formulation process and starting materials	R
Composition	R
Analytical methodology (AI) for post-registration monitoring	R
Physical properties: colour, odour, physical state, specific gravity, pH, formulation type, container type, explosivity, viscosity, technical characteristics	R
Corrosion characteristics and stability of formulation during storage	R
Data for assessment of health risk	
Summary	R
Toxicology	
Acute oral toxicity: TGAI and EP	R
Acute dermal toxicity: TGAI and EP	R
Primary eye irritation: TGAI and EP	R
Primary dermal irritation: TGAI and EP	R
Dermal sensitization/reporting of hypersensitivity incidents: TGAI and EP	R
Mutagenicity (gene mutation in microbes and mammalian cell systems and chromosome aberration): TGAI	R
Medical data, available information: TGAI and EP	R
Short-term study by appropriate route: TGAI	CR/R
Teratogenicity/developmental toxicity/one species: TGAI	CR/R

Continued

Table 3.4. *Continued*

Information, test or study	R or CR
Long-term toxicity (chronic) and carcinogenicity	CR
Multigeneration reproduction, teratogenicity (in second species), animal metabolism, neurotoxicity, immunotoxicity	CR
Occupational or bystander exposure (using the EP)	
Use description/scenario (application and post-application)	R
Passive dosimetry (mixer/loader/applicator and/or post-application) or biological monitoring	CR
Dislodgeable residues	CR
Ambient air samples	CR
Biological monitoring	CR
Dermal absorption	CR
Clothing penetration, epidemiology, package integrity	CR
Metabolism studies and residue analysis of food, feed and tobacco	
Metabolism/toxicokinetics study on animals and plants which may be directly exposed to semiochemicals through use	CR
Analytical residue methodology for food crops	CR
Crop residue data	CR
Meat, milk, poultry and egg residue data	CR
Freezer storage stability, produce quality	CR
Data for assessment of environmental risk (using EP unless otherwise specified)	
Summary	R
Effects on non-target organisms	
Birds dietary toxicity	CR
Bees: prefer EP	CR/R
Other terrestrial arthropods (crop-specific beneficial, related species): prefer EP	CR/R
Freshwater invertebrate acute toxicity: prefer EP	CR/R
Freshwater fish acute toxicity: prefer EP	CR/R
Algae: prefer EP	R
Earthworms	R
Soil microorganisms	R
Long-term laboratory or field testing on: aquatic animals, terrestrial animals, non-target plants, non-target insects	CR
Environmental fate	
Assessment based on available information	R
Experimental studies in compartments of possible concern: hydrolysis (TGAI), phototransformation on soil and/or in water (TGAI), stability in air, persistence of volatiles (TGAI), biotransformation (aerobic soil and/or aerobic aquatic, TGAI), adsorption–desorption (TGAI), leaching of each AIC from dispenser by water (EP), volatilization from dispenser, release rate (EP)	CR
Efficacy (for the EP)	
Efficacy summary	R
Description of pest problem and AI's mode of action	R
Efficacy trials of product, used as directed on the label, including reporting of adverse effects to site (e.g. phytotoxicity)	R
Sustainability considerations (compatibility with integrated pest management; contribution to risk reduction)	R

R, required data, surrogate data or a rationale to waive data; CR, required only under certain conditions; CR/R, the information is required only under certain conditions in Canada, the USA and Switzerland but is required in the EU, with the understanding that there is an appropriate basis for waiver rationale.

with biologically based pest management and were therefore slow to appreciate the need to adapt the regulations to make them appropriate for the features and characteristics of biopesticides. The questions, information and procedures required to effectively evaluate the environmental fate and behaviour of a living, microbial biopesticide are very different from those for a chemical pesticide, for example. Often the data required by regulators during this period were excessive and incurred large financial costs on the applicant that were out of proportion to the profit that was set to be made from biopesticide sales. Evaluating biopesticides using an inappropriate set of regulations has had the unintended consequence of deterring companies from commercializing biopesticide products. However, when the regulations are tailored appropriately, it is possible to achieve effective governance that does not impede the commercial development of products. Thus regulators have made a consistent effort in the last 5 years to adapt the data requirements to make them suitable for biopesticides. Usually the amount of data required now is much less than that for a conventional synthetic pesticide. Of course, the drivers behind regulation vary according to the circumstances of each country. As we have seen in the case of Cuba, the collapse of the country's conventional, industrial agriculture forced the authorities to be proactive in getting biopesticides into the hands of growers as part of the drive to develop an alternative agricultural system. Countries in the Global North clearly do not yet face the perilous and acute situation suffered by the Cubans. But policies put in place in the EU, North America and elsewhere to restrict the use of conventional pesticides, combined with the realization that agricultural production has to increase significantly in the next 20 years and at the same time become more sustainable, means that governments are having to think long and hard about biopesticides and other alternative agents as part of a new effort with IPM.

There are a number of good reasons why the sale and use of biopesticides need to be regulated. First, if a product is being sold with the express intention of controlling a pest, then it goes without saying that its effectiveness needs to be demonstrated to the people who will buy it. The EU requires that efficacy be proven as a condition for official registration, whereas other countries such as the USA tend to let the market decide whether a product is effective or not. Biopesticides are valuable components of IPM but they often have lower levels of efficacy than conventional pesticides, and hence regulators are now tending to alter their data requirements so that a biopesticide does not have to show 100% efficacy in trials in order to be granted approval. Biopesticides respond to environmental conditions significantly more than conventional pesticides and thus the level of pest control is likely to vary at different times. Systems of using biopesticides need to be developed that allow for some performance variation but which still contribute in a meaningful way to pest control. This also needs to be reflected in the advice and information given by biopesticide companies to their customers so that expectations are reasonably managed. Obviously, if a biopesticide product is not effective then farmers and growers are unlikely to use it, but customer confidence can also be damaged if a product is 'over-sold'.

Second, regulatory authorities need to determine whether a biopesticide has potential to cause negative effects on humans and the environment. They then need to decide whether any risk is associated with the biopesticide and, if so, is it acceptable? This brings us on to the third, related reason for regulation: the need to be able to characterize biopesticide products. Regulators have to ensure that biopesticide companies are selling products that contain the ingredients that are stated on the product label and are free from harmful contaminants.

So what potential hazards could be presented by biopesticides? To begin with, if the biopesticide had a toxic mode of action, it could offer a toxicity hazard to people (farm workers, production plant operatives or the general public) as well as to animals, plants and other non-target organisms. For microbial biopesticides, there are three additional potential hazards (Cook *et al.*, 1996): (i) ecological displacement of non-target microorganisms; (ii) causing an infection in non-target organisms; and (iii) causing an allergic reaction in humans or other animals. Points (i) and (ii) result from the ability of a microbial natural enemy to grow and reproduce in its environment and are features that are important for the effectiveness of many microbial biopesticides against their target pest (Cook *et al.*, 1996).

At least as far as the OECD countries are concerned (and also probably for most other countries), the regulations are set up so that biopesticides are only being approved for use if they pose minimal or zero risk. For example, the basic requirements of most OECD countries for microbial biopesticides are that:

> the microorganism and its metabolites pose no concerns of pathogenicity or toxicity to mammals and other non-target organisms which will likely be exposed to the microbial product; the microorganism does not produce a known genotoxin; all additives in the microbial manufacturing product and in end-use formulations are of low toxicity and suggest little potential for human health or environmental hazard.
>
> (OECD, 2003)

Thus, of the 218 biopesticide active substances approved by the US EPA, 206 are classed as having no expected health risk to humans. Of the remainder: nine are classed as having potential to cause eye and/or skin irritation; one is classed as having dermal and oral toxicity at high concentrations; one is slightly toxic; and one is classed as having a 'very small potential risk to human health' (United States Environmental Protection Agency, 2010). These evaluations are based on experimental investigations using toxicological and eco-toxicological tests, host range testing, as well as basic information about the biopesticide agent, such as its mode of action, which are combined into an overall risk assessment.

The risk (i.e. probability) of a negative effect happening is a combination of the hazard and the exposure. With conventional pesticides, products that are toxic to humans and other non-target animals are still authorized for use if it can be shown that the risk can be managed effectively by controlling the level of exposure, for example through the application method, by using protective equipment, by formulating and packaging the product and so forth. Because most biopesticides are deemed by regulators to present low or minimal hazard, this has the considerable advantage that less stringent precautions are required over exposure. For example, because many biopesticides leave no chemical residue, they are granted a short 'harvest interval' (the time allowed between when the product is applied and the crop is harvested). This means that pests can continue to be managed right up until the last moment. Similarly, because biopesticides normally have zero mammalian toxicity, farm workers and other operatives can handle the crop a very short time after it has been treated. This can be a vital consideration for crops grown under protection, which tend to require a high level of maintenance and handling. With conventional pesticides, workers have to be excluded from the crop for many hours or even days after spraying. The main hazard from authorized biopesticides is the development of an allergy in people working in microbial biopesticide production plants who could be regularly exposed to high amounts of microbial spores or farm workers who handle the product in concentrated form (Cook *et al.*, 1996). Only a very small fraction of microbial species produce spores that cause an allergic reaction (Latge and Paris, 1991) but nevertheless it makes sense, as part of good operating practice, to prevent exposure during production or biopesticide application by using remote-controlled machinery and by wearing personal protective equipment during spray application, although often only basic protective equipment is required.

This is not to say that all agents that could be used as a biopesticide present no hazard. Just because something is 'natural' does not mean that it is safe. It is important, therefore, that biopesticides continue to be evaluated for their human and environmental safety. Let us consider microbial natural enemies as an example. We have seen earlier in this book that some plant pathogens, such as *Aspergillus* and *Claviceps* species, produce metabolites that are toxic to people and animals. Obviously, these species would never be considered as bioherbicides according to OECD guidelines as there is a risk of the toxins entering the food chain from fungus-infected weed material accidentally harvested with the crop. Likewise, there are members of the bacterial genus *Rickettsiella* that are pathogens of arthropods (Larsson, 1978). Although they are not related antigenically to rickettsial pathogens of vertebrates, some can cause infection in the lungs of mammals if inhaled (Tanada and Kaya, 1993). Consequently, these bacteria would not be considered suitable for use as bioinsecticides as they pose a danger to spray operatives and other farm workers. There is an important role here for using basic scientific knowledge and common sense to select candidate agents for biopesticide development that are known to offer minimal hazard. From our knowledge of the biopesticides industry, there is a clear drive to identify and commercialize agents that are known to have minimal hazard.

Safety considerations for microbial biopesticides

We do not have space in this book to provide a detailed account of the potential health and safety issues for all microorganisms used as biopesticides.

We have made reference in previous sections to the safety of some agents, such as Bt and entomopathogenic viruses. For more information the reader is referred to Burges (1981) and the EU REBECA (Regulation of Environmental Biological Control Agents) website (www.rebeca-net.de). Here we will focus on a small number of illustrative examples.

We have just seen that the biopesticides that are approved for use are considered to present no hazard to human health. It is important at this juncture to put concerns about the safety of microorganisms used as biopesticides into context. Humans coexist with thousands of different microbial species as part of normal life. These include microbial natural enemies of arthropods, plant pathogens and weeds. The vast majority of these bacteria, fungi, oomycetes, viruses and protozoans are harmless to us. This is not to say that all candidate microbial agents are safe. Regulators need hard evidence on the risks of microbial biopesticides, but the problem for many biopesticide companies is that the costs of generating the required data are very high compared with the market size. There is relevant information on microorganism safety from the scientific literature that is in the public domain and could be used to support a data registration package, but it is highly specialized and can be difficult for the companies – which tend to be small enterprises – to track down and synthesize. For this reason, researchers at academic institutions can play a very important role by writing independent systematic literature reviews on microorganism safety and putting them into the public domain so that regulators, commercial operations and others in the policy network can use them. Historical experience with microbial biopesticides is also important. For example, commercial products of Bt have been authorized in the USA since the early 1960s, and a large number of studies since then have shown them to have a first-rate safety record (Siegel, 2001). This should give confidence to regulators when asked to evaluate new Bt products.

The main concern about the risks of microbial biopesticides to human health is whether the microorganism produces compounds that are toxic. Fungi, for example, produce a wide range of bioactive metabolites, some of which are toxic, although others are used as pharmaceutical medicines. Metabolites that are toxic have been characterized for a number of fungi used as microbial biopesticides, and the safety concern is that these could cause harm to people if they enter the food chain. Strasser *et al.* (2000) reviewed the toxic metabolites produced by species of the entomopathogenic fungi *Metarhizium*, *Tolypocladium* and *Beauveria*. Some of the compounds are important for the pathogenicity of the fungi to insects, although the function of others remains to be elucidated. The levels of these metabolites produced *in vivo* are usually much lower than those produced in laboratory culture, and studies have shown that negligible amounts of metabolite are released into the environment from formulated biopesticide product or from fungus-killed insects. The authors concluded that the use of fungal bioinsecticides would not result in harmful toxin levels in the environment and posed no serious risk to humans. Similarly, Zimmerman reviewed the safety of *B. bassiana* and *B. brongniartii* (Zimmerman, 2007a) and *M. anisopliae* (Zimmerman, 2007b) including comprehensive analysis of biological properties (including toxin production), analytical methods for residues, fate and behaviour in the environment, effects on non-target organisms, vertebrates, mammals and human health. On the basis of the available published information he concluded that these fungi should be considered safe to use as biopesticides. The US EPA concurs, and classify the *M. anisopliae* and *B. bassiana* strains registered as biopesticides in the USA as presenting no toxicity risk to human health. Indeed, there could even be benefits for human health: *Beauveria*, in the form of its teleomorph (sexually reproducing phase) *Cordyceps sinensis*, collected from naturally infected ghost moth caterpillars, has been used as a traditional medicine in China and Tibet for hundreds of years (Muller-Kogler, 1965, cited in Zimmerman, 2007a). The fungus is highly valuable and is a significant source of export income to the region.

The other concern for microbial biopesticides is their risk to wildlife. Effective methodologies need to be in place to determine the impact on non-target organisms. Such methodologies should be informed by ecological theory, including insights made in recent years in community ecology and invasion biology (Pearson and Callaway, 2003, 2005). Impacts on non-target organisms can be direct or indirect (e.g. competition between introduced and indigenous natural enemies). A microbial biopesticide with a high level of selectivity means that unwanted direct effects on non-target organisms are likely to be rare. However, even host-specific biological control agents can have impacts on non-target organisms through indirect effects (Pearson and Callaway, 2005).

Augmentative applications of microbial biopesticides use natural enemies that are endemic, i.e. that already occur naturally in the country or region of use. They are not aimed at permanent establishment and the population of the released agent is expected to decline to background levels post-application. Therefore, any negative effects on non-target species should be temporary. Cook *et al.* (1996) state that, 'there is nothing inherent in the strategy itself (inoculative, augmentative, or inundative) that raises a safety issue'. Practical experience with agents such as entomopathogenic fungi used for augmentation biological control backs this up, with no detectable detrimental environmental impact (Goettel *et al.*, 2001; Vestergaard *et al.*, 2003). Such experience has an important bearing on the risk evaluation of new products, but – as outlined above – this is not to say that evaluation of new products is not required. To start with, augmentative applications alter interaction in space and time between the microbial agent, its hosts and the environment (Jackson, 2003). If microbial biopesticides become used more widely, then the amount of environmental perturbation might increase. Biopesticide manufacturers are under commercial pressures to develop products with a relatively wide host range and this increases the risk of a negative environmental effect. Potentially, there could be unintended effects, for example, on the diversity and function of other microbial natural enemies. Host range evaluation is important for microbial biopesticides, although procedures here have been criticized for concentrating on the physiological host range of agents (i.e. the potential host range as determined through laboratory bioassays) at the expense of studies of the ecological host range (i.e. the actual host range in the agro-ecosystem, which tends to be much narrower) (Jaronski et al., 2003).

Since classical control is based on the deliberate introduction of a non-indigenous natural enemy with the aim of permanent establishment, determination of host specificity is critical to ensure that agents released do not have negative effects on non-target organisms. There are now well-established systems for risk assessment and host range testing (Andersen et al., 2005) that tie in to the FAO code of conduct on the import and release of biological control agents (FAO, 1996). But it has been argued that there is still a lack of long-term, quantitative and objective monitoring of classical control programmes (Thomas and Reid, 2007). This may be because few apparent problems have been encountered with classical control (van Lenteren et al., 2006; Hajek and Delalibera, 2010). However, where pre-release risk evaluation procedures are inadequate or ignored, environmental damage can occur. It should not be forgotten that early introductions of alien generalist predators, such as cane toads in Australia and coccinellids in Hawaii, were done without proper consideration of the risks and with a poor understanding of ecological principles, resulting in unacceptable environmental consequences (Thomas and Willis, 1998; Barratt et al., 2010). And a prominent recent example in Western Europe concerns the harlequin ladybird, *Harmonia axyridis*. This species is native to Asia and has been used as a control agent of aphids in glasshouses in Europe and North America. It has been intentionally introduced in nine European countries since 1982 (Brown et al., 2008), although a retrospective analysis identified it as having high environmental risk (van Lenteren et al., 2003) and thus it should not have been released (van Lenteren et al., 2008). It is now established in 13 European countries from Denmark to southern France and is predicted to spread further. It is able to outcompete native ladybirds, will predate on some beneficial insects, and there is evidence that it has a significant negative impact on other native arthropod species. This episode has undoubtedly cast a shadow over classical biological control in Europe. This is unfortunate, because it remains the only method for permanent ecological management of many alien invasive species. It has been proposed that legislation is enacted within the EU in line with the Convention on Biological Diversity to enable releases of classical control agents based on EPPO (European and Mediterranean Plant Protection Organization) standards (Sheppard et al., 2006). In the future there is likely to be an even greater drive to reconcile biocontrol efficacy with biosafety, with the focus being on selecting biological control agents that are effective at controlling the target pest but which are highly selective and present no risk to non-target species (Barratt et al., 2010).

There is a related issue that crosses the divide between classical and augmentative biocontrol, namely whether an entity of a microbial natural enemy intended for use in augmentation biocontrol is 'endemic' or not. For example, many fungal species that have been classified taxonomically on the basis of morphological criteria are said to have worldwide distributions. However, studies using molecular tools indicate that in reality these individual 'morphological' species may consist of an assemblage of genetically distinct 'cryptic' or 'hidden' species, each with its own characteristics and with differing geographical distributions (Desprez-Loustau *et al.*, 2007). If this were to apply to a fungal species being considered as a microbial biopesticide, there could be a chance of unknowingly introducing a non-native, 'cryptic' species to a new country. There is a clear role here for research on the diversity and biogeography of microbial natural enemies to underpin environmental risk evaluation of microbial biopesticides. For example, a molecular phylogenetic study of the entomopathogenic fungus *B. bassiana* showed that a strain of the fungus isolated in the USA, and now used in a commercial biopesticide sold in the USA and Europe, is a member of a cryptic species that is indigenous to North America and Europe (Rehner and Buckley, 2005). Such basic studies on microbial phylogeny have an important role to play in the future regulation of microbial biopesticides.

Safety considerations for botanicals and semiochemicals

If a botanical pesticide comprises just a single active ingredient then its registration data requirements are relatively straightforward and it can be assessed in much the same way as a conventional chemical pesticide. Reduced data requirements are likely if there is well-established evidence, e.g. from the scientific literature, European pharmacopoeia, etc., that the compound is non-toxic to mammals and other non-targets and presents minimal risk. Problems are likely to arise, however, if the botanical consists of a complex plant extract that is difficult to characterize and is likely to vary between production batches.

Semiochemicals are naturally occurring non-toxic molecular messengers that work by chemically binding to receptor molecules that are located on the cell membranes of the receiving organism. They are target specific: if an organism does not possess the receptors for a particular semiochemical (e.g. in the case of a human exposed to insect sex pheromones) then the semiochemical will not bind to its cells. Therefore, there are sound biological reasons for government regulators approaching semiochemicals intended for use as biopesticides from the starting position that they are unlikely to be hazardous to non-target organisms. The OECD regards semiochemicals used for arthropod control as presenting minimal hazard to people or the environment, and it considers SCLPs, which form the majority of semiochemical-based biopesticides, as particularly safe (OECD, 2001). This assessment was based on the fact that they are non-toxic to humans, the rates at which they are applied in the field are very low and are typically similar to levels that occur naturally, and that they dissipate and degrade rapidly in the environment. Moreover, many are applied in passive dispensers that prevent direct contact between the semiochemical concentrate and humans or wildlife. Most organisms are able to degrade SCLPs by normal metabolic processes, and the OECD considers that they 'should present no problems with their normal physiology' if contamination does occur (OECD, 2001). The OECD recommends that the high safety factors:

> justify substantial reductions in health and environmental data requirements, especially for SCLPs, a well-defined chemical group for which considerable data are available. Also for other classes of semiochemicals, it may be justified to waive certain required studies if the registrant can provide an adequate rationale.
> (OECD, 2001)

References

Alabouvette, C., Olivain, C. and Steinberg, C. (2006) Biological control of plant diseases: the European situation. *European Journal of Plant Pathology* 114, 329–341.

Andersen, M.C., Ewald, M. and Northcott, J. (2005) Risk analysis and management decisions for weed biological control agents: ecological theory and modelling results. *Biological Control* 35, 330–337.

Anon. (2006) The Greening Waipara Project. *Bio-Protection* issue 2, November. Available at: http://bioprotection.org.nz/system/files/Greening%20Waipara%20Newsletter%20No%202.pdf (accessed 13 July 2010).

Ash, G.J. (2010) The science, art and business of successful bioherbicides. *Biological Control* 52, 230–240.

Asser-Kaiser, S., Fritsch, E., Undorf-Spahn, K., Kienzle, J., Eberle, K., Gund, N.A., Reineke, A., Zebitz, C.P.W., Heckel, D.G., Huber, J. and Jehle, J.A. (2007) Rapid emergence of baculovirus resistance in codling moth due to dominant, sexlinked inheritance. *Science* 318, 1916–1917.

Atawodi, S.E. and Atawodi, J.C. (2009) *Azadirachta indica* (neem): a plant of multiple biological and pharmacological activities. *Phytochemistry Reviews* 8, 601–620.

Bailey, K.L., Boyetchko, S.M. and Langle, T. (2010) Social and economic drivers shaping the future of biological control: a Canadian perspective on the factors affecting the development and use of microbial biopesticides. *Biological Control* 52, 221–229.

Bale, J.S., van Lenteren, J.C. and Bigler, F. (2008) Biological control and sustainable food production. *Philosophical Transactions of the Royal Society B* 363, 761–776.

Barratt, B.I.P., Howarth, F.G., Withers, T.M., Kean, J.M. and Ridley, G.S. (2010) Progress in risk assessment for classical biological control. *Biological Control* 52, 245–254.

Baxter, S.W., Chen, M., Dawson, A., Zhao, J.-Z., Vogel, H., Shelton, A.M., Heckel, D.G. and Jiggins, C.D. (2010) Mis-spliced transcripts of nicotinic acetylcholine receptor α6 are associated with field evolved Spinosad resistance in *Plutella xylostella* (L.). *PLoS Genetics* 6, e1000802.

Bellotti, A.C., Smith, L. and Lapointe, S.L. (1999) Recent advances in cassava pest management. *Annual Review of Entomology* 44, 343–370.

Bellows, T.S. and Fisher, T.W. (eds) (1999) *Handbook of Biological Control*. Academic Press, San Diego, California.

Bending, G.D., Aspray, T.A. and Whipps, J.M. (2008) Significance of microbial interactions in the mycorrhizosphere. *Advances in Applied Microbiology* 60, 97–132.

Berg, G. (2009) Plant–microbe interactions promoting plant growth and health: perspectives for controlled use of microorganisms in agriculture. *Applied Microbiology and Biotechnology* 84, 11–18.

Berndt, L.A. and Wratten, S.D. (2005) Effects of alyssum flowers on the longevity, fecundity and sex ratio of the leaf roller parasitoid *Dolichogenidae tasmanica*. *Biological Control* 32, 65–69.

Berndt, L.A., Wratten, S.D. and Scarratt, S.L. (2006) The influence of floral resource subsidies on parasitism rates of leafrollers (Lepidoptera: Totricidae) in New Zealand vineyards. *Biological Control* 37, 50–55.

Bielza, P., Quinto, V., Contreras, J., Torné, M., Martin, A. and Espinosa, P.J. (2007) Resistance to Spinosad in the western flower thrips, *Frankliniella occidentalis* (Pergande), in greenhouses of south-eastern Spain. *Pest Management Science* 63, 682–687.

Bond, R.P.M., Boyce, C.B.C., Rogoff, M.H. and Shieh, T.R. (1971) The thermostable exotoxin of *Bacillus thuringiensis*. In: Burges, H.D. and Hussey, N.W. (eds) *Microbial Control of Insects and Mites*. Academic Press, London, pp. 275–303.

Brar, S.K., Verma, M., Tyagi, R.D. and Valero, J.R. (2006) Recent advances in downstream processing and formulations of *Bacillus thuringiensis* based biopesticides. *Process Biochemistry* 41, 323–342.

Brown, P.M.J., Adriaens, T., Bathon, H., Cuppen, J., Goldarazena, A., Hagg, T., Kenis, M., Klausnitzer, B.E.M., Kovar, I., Loomans, A.J.M., Majerus, M.E.N., Nedved, O., Perdersen, J., Rabitsch, W., Roy, H.E., Ternois, V., Zakharov, I.A. and Roy, D.B. (2008) *Harmonia axyridis* in Europe: spread and distribution of a non-native coccinellid. *BioControl* 53, 5–21.

Burges, H.D. (1981) Safety, safety testing and quality control of microbial pesticides. In: Burges, H.D. (ed.) *Microbial Control of Pests and Plant Diseases 1970–1980*. Academic Press, London, pp. 737–767.

Chandler, D., Davidson, G. and Jacobson, R.J. (2005) Laboratory and glasshouse evaluation of entomopathogenic fungi against the two-spotted spider mite, *Tetranychus urticae* (Acari: Tetranychidae) on tomato, *Lycopersicon esculentum*. *Biocontrol Science and Technology* 15, 37–54.

Chaney, W.E. (1998) Biological control of aphids in lettuce using in-field insectaries. In: Pickett, C.H. and Bugg, R.L. (eds) *Enhancing Biological Control: Habitat Management to Promote Natural Enemies of Agricultural Pests*. University of California Press, Berkeley, California, pp. 73–83.

Charudattan, R. (1990) The mycoherbicide approach with plant pathogens. In: te Beest, D.O. (ed.) *Microbial Control of Weeds*. Chapman and Hall, London, pp. 24–57.

Charudattan, R. (2001) Biological control of weeds by means of plant pathogens: significance for integrated weed management in modern agro-ecology. *BioControl* 46, 229–260.

Choudhary, D.K. and Johri, B.N. (2009) Interactions of *Bacillus* spp. and plants – with special reference to induced systemic resistance (ISR). *Microbiological Research* 164, 493–513.

Cloyd, R.A. and Chiasson, H. (2007) Activity of an essential oil derived from *Chenopodium ambrosioides* on greenhouse insect pests. *Journal of Economic Entomology* 100, 459–466.

Cloyd, R.A., Galle, C.L., Keith, S.R., Kalscheur, N.A. and Kemp, K.E. (2009) Effect of commercially available plant-derived essential oil products on arthropod pests. *Journal of Economic Entomology* 102, 1567–1579.

Collier, T. and van Steenwyck, R. (2004) A critical evaluation of augmentative biological control. *Biological Control* 31, 245–256.

Cook, R.J., Bruckart, W.L., Coulson, J.R., Goettel, M.S., Humber, R.A., Lumsden, R.D., Maddox, J.V., McManus, M.L., Moore, L., Meyer, S.F., Quimby, P.C., Stack, J.P. and Vaughn, J.L. (1996) Safety of microorganisms intended for pest and plant disease control: a framework for scientific evaluation. *Biological Control* 7, 333–351.

Cook, S.M., Khan, Z.R. and Pickett, J.A. (2007) The use of push–pull strategies in integrated pest management. *Annual Review of Entomology* 52, 375–400.

Cooper, S.G., Douches, D.S. and Grafius, E.J. (2004) Combining genetic engineering and traditional breeding to provide elevated resistance in potatoes to Colorado potato beetle. *Entomologia Experimentalis et Applicata* 112, 37–46.

Copping, L.G. (2004) *The Manual of Biocontrol Agents*. British Crop Protection Council, Farnham, UK, 752 pp.

Dedej, S., Delaplane, K.S. and Scherm, H. (2004) Effectiveness of honey bees in delivering the biocontrol agent *Bacillus subtilis* to blueberry flowers to suppress mummy berry disease. *Biological Control* 31, 422–427.

Delalibera, I. Jr, Sosa Gomez, D.R., de Moraes, G.J., Alencar, J.A. and Farias Araujo, W. (1992) Infection of *Mononychellus tanajoa* (Acari: Tetranychidae) by the fungus *Neozygites* sp. (Entomophthorales) in northeastern Brazil. *Florida Entomologist* 75, 145–147.

de Maagd, R.A., Bosch, D. and Stiekema, W. (1999) *Bacillus thuringiensis* toxin-mediated insect resistance in plants. *Trends in Plant Science* 4, 9–13.

Desprez-Loustau, M.-L., Robin, C., Buee, M., Courtecuisse, R., Garbaye, J., Suffert, F., Sache, I. and Rizzo, D.M. (2007) The fungal dimension of biological invasions. *Trends in Ecology and Evolution* 22, 472–480.

Dick, G.L., Buschman, L.L. and Ramoska, W.A. (1992) Description of a species of *Neozygites* infecting *Oligonychus pratensis* in the western Great Plains of the United States. *Mycologia* 84, 729–738.

Dong, L.Q. and Zhang, K.Q. (2006) Microbial control of plant-parasitic nematodes: a five party interaction. *Plant and Soil* 288, 31–45.

Dyer, G.A., Serratos-Hernandez, J.A., Perales, H.R., Gepts, P., Pineyro-Nelson, A., Chavez, A., Salinas-Arreortua, N., Yunez-Naude, A., Taylor, J.E. and Alvarez-Buylla, E.R. (2009) Dispersal of transgenes through maize seed systems in Mexico. *PLoS One* 4, e5734.

Eilenberg, J., Hajek, A. and Lomer, C. (2001) Suggestions for unifying the terminology in biological control. *BioControl* 46, 387–400.

El-Sayed, A.M., Suckling, D.M., Byers, J.A., Jang, E.B. and Wearing, C.H. (2009) Potential of 'lure and kill' in long-term pest management and eradication of invasive species. *Journal of Economic Entomology* 102, 815–835.

EPPO (2002) List of biological control agents widely used in the EPPO region. *EPPO Bulletin* 32, 447–461.

FAO (1996) *Code of Conduct for the Import and Release of Exotic Biological Control Agents. International Standard for Phytosanitary Measures No. 3*. Secretariat of the International Plant Protection Convention, Food and Agriculture Organization of the United Nations, Rome.

FAO (2007) *The Future of Biopesticides in Desert Locust Management. Report of the Internatonal Workshop, Saly, Senegal, 12–15 February 2007*. Food and Agriculture Organization of the United Nations, Rome.

Faria, M.R. and Wraight, S.P. (2007) Mycoinsecticides and mycoacaricides: a comprehensive list with worldwide coverage and international classification of formulation types. *Biological Control* 43, 237–256.

Fielder, A.K., Landis, D.A. and Wratten, S.D. (2008) Maximizing ecosystem services from conservation biological control: the role of habitat management. *Biological Control* 45, 254–271.

Freckleton, R.P., Sutherland, A.R. and Watkinson, A.R. (2003) Deciding the future of GM crops in Europe. *Science* 302, 994–996.

Gelernter, W.D. (2005) Biological control products in a changing landscape. In: *The BCPC International Congress Proceedings: 2005*, Vol. 1. British Crop Protection Council, Alton, UK, pp. 293–300.

Georgis, R., Koppenhofer, A.M., Lacey, L.A., Belair, G., Duncan, L.W., Grewal, P.S., Samish, M., Tan, L., Torr, P. and van Tol, R.W.H.M. (2006) Successes and failures in the use of parasitic nematodes for pest control. *Biological Control* 38, 103–123.

Gill, S.S., Cowles, E.A. and Pietrantonio, P.V. (1992) The mode of action of *Bacillus thuringiensis* endotoxins. *Annual Review of Entomology* 37, 615–636.

Goettel, M.S., Hajek, A.E., Siegel, J.P. and Evans, H.C. (2001) Safety of fungal biocontrol agents. In: Butt, T.M., Jackson, C. and Magan, N. (eds) *Fungi as Biocontrol Agents: Progress, Problems and Potential*. CAB International, Wallingford, UK, pp. 347–376.

Gosling, P., Hodge, A., Goodlass, G. and Bending, G.D. (2006) Arbuscular mycorrhizal fungi and organic farming. *Agriculture, Ecosystems & Environment* 113, 17–35.

Groner, A. (1990) Safety to nontarget invertebrates of baculoviruses. In: Laird, M., Lacey, L.A. and Davidson, E.W. (eds) *Safety of Microbial Insecticides*. CRC Press, Boca Raton, Florida, pp. 135–147.

Grzywacz, D., Rossbach, A., Rauf, A., Russell, D.A., Srinivasan, R. and Shelton, A.M. (2010) Current control methods for diamondback moth and other brassica insect pests and the prospects for improved management with lepidopteran-resistant Bt vegetable crops in Asia and Africa. *Crop Protection* 29, 68–79.

Guleria, S. and Tiku, A.K. (2009) Botanicals in pest management: current status and future perspectives. In: Peshin, R. and Dhawan, A.K. (eds) *Integrated Pest Management: Innovation-Development Process*, Vol. 1. Springer, Dordrecht, the Netherlands, pp. 317–329.

Haas, D. and Défago, G. (2005) Biological control of soil-borne pathogens by fluorescent pseudomonads. *Nature Reviews Microbiology* 3, 307–319.

Hajek, A. (2004) *Natural Enemies: an Introduction to Biological Control*. Cambridge University Press, Cambridge.

Hajek, A.E. and Delalibera, I. (2010) Fungal pathogens as classical biological control agents against arthropods. *BioControl* 55, 147–158.

Hajek, A.E., McManus, M.L. and Delalibera, I. (2007) A review of introductions of pathogens and nematodes for classical biological control of insects and mites. *Biological Control* 41, 1–13.

Hallett, S.G. (2005) Where are the bioherbicides? *Weed Science* 53, 404–415.

Hara, A.H., Kaya, H.K., Gaugler, R., Lebeck, L.M. and Mello, C.L. (1993) Entomopathogenic nematodes for biological control of the leafminer *Liriomyza trifolii* (Dipt.: Agromyzidae). *Entomophaga* 38, 359–369.

Hare, J.D. (1990) Ecology and management of the Colorado potato beetle. *Annual Review of Entomology* 35, 81–100.

Harman, G.E. (2005) Overview of mechanisms and uses of *Trichoderma* spp. *Phytopathology* 96, 190–194.

Hawksworth, D.L., Kirk, P.M., Sutton, B.C. and Pegler, D.N. (1995) *Ainsworth and Bisby's Dictionary of the Fungi*, 8th edn. CAB International, Wallingford, UK.

He, L., Gao, X., Wang, J., Zhao, Z. and Liu, N. (2009) Genetic analysis of abamectin resistance in *Tetranychus cinnabarinus*. *Pesticide Biochemistry and Physiology* 95, 147–151.

Herron, G.A. and James, T.M. (2005) Monitoring insecticide resistance in Australian *Frankliniella occidentalis* Pergande (Thysanoptera: Thripidae) detects fipronil and Spinosad resistance. *Australian Journal of Entomology* 44, 299–303.

Hollingsworth, R.G., Steinkraus, D.C. and McNew, R.W. (1995) Sampling to predict fungal epizootics in cotton aphids (Homoptera: Aphididae). *Environmental Entomology* 24, 1414–1421.

Hountondji, F.C.C., Hanna, R., Cherry, A.J., Sabelis, M.W., Agboton, B. and Korie, S. (2007) Scaling-up tests on virulence of the cassava green mite fungal pathogen *Neozygites tanajoae* (Entomophthorales: Neozygitaceae) under controlled conditions: first observations at the population level. *Experimental and Applied Acarology* 41, 153–168.

Huber, J. (1986) Use of baculoviruses in pest management programs. In: Granados, R.R. and Federici, B.A. (eds) *The Biology of Baculoviruses*. Vol. II. *Practical Application for Insect Control*. CRC Press, Boca Raton, Florida, pp. 181–202.

Huger, A.M. (2005) The *Oryctes* virus: its detection, identification, and implementation in biological control of the coconut palm rhinoceros beetle, *Oryctes rhinoceros* (Coleoptera: Scarabaeidae). *Journal of Invertebrate Pathology* 89, 78–84.

International Service for the Acquisition of Agri-Biotech Applications (2010) Global status of commercialised biotech/GM crops 2009. Available at: http://www.isaaa.org/ (accessed 29 June 2010).

Iwata, M. (2001) Probenazole – a plant defence activator. *Pesticide Outlook* 12, 28–31.

Jackson, T.A. (2003) Environmental safety of inundative application of a naturally occurring biocontrol agent, *Serratia entomophila*. In: Hokkanen, H.M.T. and Hajek, A.E. (eds) *Environmental Impacts of Microbial Insecticides*. Kluwer Academic Publishers, Dordrecht, the Netherlands, pp. 169–176.

Jacobson, R.J., Chandler, D., Fenlon, J. and Russell, K.M. (2001) Compatibility of *Beauveria bassiana* (Balsamo) Vuillemin with *Amblyseius cucumeris* Oudemans (Acarina: Phytoseiidae) to control *Frankliniella occidentalis* Pergande (Thysanoptera: Thripidae) on cucumber plants. *Biocontrol Science and Technology* 11, 381–400.

James, D.G., Castle, S.C., Grasswitz, T. and Reyna, V. (2005) Using synthetic herbivore-induced plant volatiles to enhance conservation biological control: field experiments in hops and grapes. In: Hoddle, M.S. (ed.) *Proceedings of Second International Symposium on Biological Control of Arthropods, Davos, Switzerland, 12–16 September 2005*. USDA Forest Service, Morgantown, West Virginia, FHTET Publication 2005-08, pp. 192–205.

Jansky, S.H., Simon, R. and Spooner, D.M. (2009) A test of taxonomic predictivity: resistance to the Colorado potato beetle in wild relatives of cultivated potato. *Journal of Economic Entomology* 102, 422–431.

Jaronski, S.T., Goettel, M.S. and Lomer, C.J. (2003) Regulatory requirements for ecotoxicological assessments of microbial insecticides – how relevant are they? In: Hokkanen, H.M.T. and Hajek, A.E. (eds) *Environmental Impacts of Microbial Insecticides*. Kluwer Academic Publishers, Dordrecht, the Netherlands, pp. 237–260.

Jehle, J.A. (2008) The future of *Cydia pomonella* granulovirus in biological control of codling moth. In: Boos, M. (ed.) *Ecofruit – 13th International Conference on Cultivation Technique and Phytopathological Problems in Organic Fruit-Growing: Proceedings to the Conference from 18–20 February 2008 at Weinsberg/Germany*. Association for the Promotion of Organic Fruit Growing, Weinsberg, Germany, pp. 265–270. Available at: http://www.ecofruit.net/proceedings-2008.html (accessed 13 July 2010).

Jonsson, M., Wratten, S.D., Landis, D.A. and Gurr, G.M. (2008) Recent advances in conservation biological control of arthropods by arthropods. *Biological Control* 45, 172–175.

Kabaluk, J.T. and Gazdik, K. (2007) *Directory of Microbial Pesticides for Agricultural Crops in OECD Countries, 2007*. Agriculture and Agri-Food Canada/Agriculture et Agroalimentaire Canada, Agassiz, British Columbia, 212 pp.

Kenney, D.S. (1986) DeVine – the way it was developed – an industrialist's view. *Weed Science* 34(Suppl. 1), 15–16.

Kerry, B.R. (2000) Rhizosphere interactions and the exploitation of microbial agents for the biological control of plant-parasitic nematodes. *Annual Review of Phytopathology* 38, 423–441.

Khan, Z.R., James, D.G., Midega, C.A.O. and Pickett, J.A. (2008) Chemical ecology and conservation biological control. *Biological Control* 45, 210–224.

Kiss, L., Russell, J.C., Szentiványi, O., Xu, X. and Jeffries, P. (2004) Biology and biocontrol potential of *Ampelomyces* mycoparasites, natural antagonists of powdery mildew fungi. *Biocontrol Science and Technology* 14, 635–651.

Kloepper, J.W., Ryu, C.M. and Zhang, S.A. (2004) Induced systemic resistance and promotion of plant growth by *Bacillus* spp. *Phytopathology* 94, 1259–1266.

Koppenhofer, A.M. (2006) Nematodes. In: Lacey, L.A. and Kaya, H.K. (eds) *Field Manual of Techniques in Invertebrate Pathology: Application and Evaluation of Pathogens for Control of Insects and Other Invertebrate Pests*. Springer, Dordrecht, the Netherlands, pp. 283–301.

Koul, O. (2008) Phytochemicals and insect control: an antifeedant approach. *Critical Reviews in Plant Sciences* 27, 1–24.

Kunert, G., Otto, S., Rose, U.S.R., Gershenzon, J. and Weisser, W.W. (2005) Alarm pheromone mediates production of winged dispersal morphs in aphids. *Ecology Letters* 8, 596–603.

Lacey, L.A. and Kaya, H.K. (eds) (2006) *Field Manual of Techniques in Invertebrate Pathology: Application and Evaluation of Pathogens for Control of Insects and Other Invertebrate Pests*. Springer, Dordrecht, the Netherlands.

Lacey, L.A. and Shapiro-Ilan, D.I. (2008) Microbial control of insect pests in temperate orchard systems: potential for incorporation into IPM. *Annual Review of Entomology* 53, 121–144.

Lacey, L.A. and Siegel, J.P. (2000) Safety and ecotoxicology of entomopathogenic bacteria. In: Charles, J.F., Delecluse, A. and Nielsen-LeRoux, C. (eds) *Entomopathogenic Bacteria: From Laboratory to Field Application*. Kluwer Academic, Dordrecht, the Netherlands, pp. 253–273.

Lacey, L.A. and Undeen, A.H. (1986) Microbial control of black flies and mosquitoes. *Annual Review of Entomology* 31, 265–296.

Larsson, R. (1978) Insects and rickettsiae. *Entomologisk Tidskrift* 99, 71–84.

Lasota, J.A. and Dybas, R.A. (1991) Avermectins, a novel class of compounds: implications for use in arthropod pest control. *Annual Review of Entomology* 36, 91–117.

Latge, J.P. and Paris, S. (1991) The fungal spore reservoir of allergens. In: Cole, G.T. and Hoch, H.C. (eds) *The Fungal Spore and Disease Initiation in Plants and Animals*. Plenum Press, New York, pp. 379–401.

Li, G.-P., Wu, K.-M., Gould, F., Wang, J.-K., Miao, J., Gao, X.-W. and Guo, Y.-Y. (2007) Increasing tolerance to Cry1Ac cotton from cotton bollworm, *Helicoverpa armigera*, was confirmed in Bt cotton farming area of China. *Ecological Entomology* 32, 366–375.

Li, Z., Alves, S.B., Roberts, D.W., Fan, M., Delalibera, I., Tang, J., Lopes, R.B., Faria, M. and Rangel, D.E.M. (2010) Biological control of insects in Brazil and China: history, current programs and reasons for their success using entomopathogenic fungi. *Biocontrol Science and Technology* 20, 117–136.

Lomer, C.J., Bateman, R.P., Johnson, D.L., Langewalkd, J. and Thomas, M. (2001) Biological control of locusts and grasshoppers. *Annual Review of Entomology* 46, 667–702.

Lu, Y.H., Qiu, F., Feng, H.Q., Li, H.B., Yang, Z.C., Wyckhuys, K.A.G. and Wu, K.M. (2008) Species composition and seasonal abundance of pestiferous plant bugs (Hemiptera: Miridae) on Bt cotton in China. *Crop Protection* 27, 465–472.

MacPherson, R.M. and MacRae, T.C. (2009) Evaluation of transgenic soybean exhibiting high expression of a synthetic *Bacillus thuringiensis cry1A* transgene for suppressing lepidopteran populations densities and crop injury. *Journal of Economic Entomology* 102, 1640–1648.

Marrone, P.G. (2007) Barriers to adoption of biological control agents and biological pesticides. *CAB Reviews: Perspectives in Agriculture, Veterinary Science, Nutrition and Natural Resources* 2(51). CAB International, Wallingford, UK.

Mertz, F.P. and Yao, R.C. (1990) *Saccharopolyspora spinosa* sp nov isolated from soil collected in a sugar mill rum still. *International Journal of Systematic Bacteriology* 40, 34–39.

Moscardi, F. (1999) Assessment of the application of baculoviruses for control of Lepidoptera. *Annual Review of Entomology* 44, 257–289.

NASS (2008) Agricultural Chemical Usage 2007 Field Crops Summary. http://usda.mannlib.cornell.edu/usda/current/AgriChemUsFruits/AgriChemUsFruits-05-21-2008.pdf (accessed 13 April 2010).

Nicholls, C.I., Perez, N., Vasquez, L. and Altieri, M. (2002) The development and status of biologically based integrated pest management in Cuba. *Integrated Pest Management Reviews* 7, 1–16.

Nishida, R. (2002) Sequestration of defensive substances from plants by Lepidoptera. *Annual Review of Entomology* 47, 57–92.

Nuffield Council on Bioethics (1999) Genetically modified crops: the ethical and social issues. http://www.nuffieldbioethics.org/go/ourwork/gmcrops/publication_301.html (accessed 28 November 2008).

OECD (2001) Guidance for Registration Requirements for Pheromones and other Semiochemicals used for Arthropod Pest Control. OECD Series on Pesticides No. 12. http://www.oecd.org/dataoecd/44/31/33650707.PDF (accessed 5 March 2010).

OECD (2003) Guidance for Registration Requirements for Microbial Pesticides. OECD Series on Pesticides No. 18. http://www.oecd.org/dataoecd/4/23/28888446.pdf (accessed 5 March 2010).

Pearson, D.E. and Callaway, R.M. (2003) Indirect effects of host-specific biological control agents. *Trends in Ecology and Evolution* 18, 456–461.

Pearson, D.E. and Callaway, R.M. (2005) Indirect nontarget effects of host-specific biological control agents: implications for biological control. *Biological Control* 35, 288–298.

Pell, J.K., Hannam, J.J. and Steinkraus, D.C. (2010) Conservation biological control using fungal entomopathogens. *BioControl* 55, 187–198.

Penalver, R., Vicedo, B., Salcedo, C.I. and Lopez, M.M. (1994) *Agrobacterium radiobacter* strains K84, K1026 and K84 Agr⁻ produce an antibiotic-like substance, active *in vitro* against *A. tumefaciens* and phytopathogenic *Erwinia* and *Pseudomonas* spp. *Biocontrol Science and Technology* 4, 259–267.

Pratt, J.E., Gibbs, J.N. and Webber, J.F. (1999) Registration of *Phlebiopsis gigantea* as a forest biocontrol agent in the UK: recent experience. *Biocontrol Science and Technology* 9, 113–118.

Rae, R., Verdun, C., Grewal, P.S., Robertson, J.F. and Wilson, M.J. (2007) Biological control of terrestrial molluscs using *Phasmarhabditis hermaphrodita* – progress and prospects. *Pest Management Science* 63, 1153–1164.

Raina, A.K., Kingan, T.G. and Mattoo, A.K. (1992) Chemical signals from host plant and sexual behaviour in a moth. *Science* 255, 592–594.

Reddy, G.V.P. and Guerrero, A. (2004) Interactions of insect pheromones and plant semiochemicals. *Trends in Plant Science* 9, 253–261.

Reddy, G.V.P., Cruz, Z.T. and Guerrero, A. (2009) Development of an efficient pheromone-based trapping method for the banana root borer *Cosmopolites sordidus*. *Journal of Chemical Ecology* 35, 111–117.

Rehner, S.A. and Buckley, E.P. (2005) Cryptic diversification in *Beauveria bassiana* inferred from nuclear ISTs and EF1-α phylogenies. *Mycologia* 97, 84–98.

Romeis, J., Meissle, M. and Bigler, F. (2006) Transgenic crops expressing *Bacillus thuringiensis* toxins and biological control. *Nature Biotechnology* 24, 63–71.

Rowell, B. and Bessin, R. (2005) Bt Basics for Vegetable Integrated Pest Management, ID-156. http://www.ca.uky.edu/agc/pubs/id/id156/id156.pdf (accessed April 2010).

Roy, H.E., Steinkraus, D.C., Eilenberg, J., Hajek, A.E. and Pell, J.K. (2006) Bizarre interactions and endgames: entomopathogenic fungi and their arthropod hosts. *Annual Review of Entomology* 51, 331–357.

Sato, M.E., Da Silva, M.Z., Raga, A. and de Souza, M.F. (2005) Abamectin resistance in *Tetranychus urticae* Koch (Acari: Tetranychidae): selection, cross-resistance and stability of resistance. *Neotropical Entomology* 34, 991–998.

Schmutterer, H. (1990) Properties and potentials of natural pesticides from neem tree. *Annual Review of Entomology* 35, 271–298.

Science online (2010) GM Crops, A World View. Special Online Collection: Plant Genomes. http://www.sciencemag.org/plantgenomes/map.html (accessed April 2010).

Sharma, R.R., Singh, D. and Singh, R. (2009) Biological control of postharvest diseases of fruits and vegetables by microbial antagonists: a review. *Biological Control* 50, 205–221.

Shelton, A.M., Zhao, J.-Z. and Roush, R.T. (2002) Economic, ecological, food safety and social consequences of the deployment of Bt transgenic plants. *Annual Review of Entomology* 47, 845–881.

Sheppard, A.W., Shaw, R.H. and Sforza, R. (2006) Top 20 environmental weeds for classical biological control in Europe: a review of opportunities, regulations and other barriers to adoption. *Weed Research* 46, 93–117.

Siegel, J.P. (2001) The mammalian safety of *Bacillus thuringiensis*-based insecticides. *Journal of Invertebrate Pathology* 77, 13–21.

Silverio, F.O., de Alvarenga, E.S., Moreno S.C. and Picanc, M.C. (2009) Synthesis and insecticidal activity of new pyrethroids. *Pest Management Science* 65, 900–905.

Smitley, D.R., Kennedy, G.G. and Brooks, W.M. (1986) Role of the entomogenous fungus, *Neozygites floridana*, in population declines of the twospotted mite, *Tetranychus urticae*, on field corn. *Entomologia Experimentalis et Applicata* 41, 255–264.

Spadaro, D. and Gullino, M.L. (2004) State of the art and future prospects of the biological control of postharvest diseases. *International Journal of Food Microbiology* 91, 185–194.

Sparks, T.C., Crouse, G.D. and Durst, G. (2001) Natural products and insecticides: the biology, biochemistry and quantitative structure–activity relationships of spinosyns and spinosoids. *Pest Management Science* 57, 896–905.

Strasser, H., Vey, A. and Butt, T.M. (2000) Are there any risks in using entomopathogenic fungi for pest control, with particular reference to the bioactive metabolites of *Metarhizium*, *Tolypocladium* and *Beauveria* species? *Biocontrol Science and Technology* 10, 717–735.

Szewczyk, B., Hoyis-Carvajal, L., Paluszek, M. Skrzecz, I. and de Souza, M.L. (2006) Baculoviruses – re-emerging biopesticides. *Biotechnology Advances* 24, 143–160.

Tanada, Y. and Kaya, H.K. (1993) *Insect Pathology*. Academic Press, San Diego, California.

te Beest, D.O., Yang, X.B. and Cisar, C.R. (1992) The status of biological control of weeds with fungal pathogens. *Annual Review of Phytopathology* 30, 637–657.

Thacker, J.R.M. (2002) *An Introduction to Arthropod Pest Control*. Cambridge University Press, Cambridge.

Thaler, J.S. (1999) Jasmonate-inducible plant defences cause increased parasitism of herbivores. *Nature* 399, 686–688.

Thomas, M.B. and Reid, A.M. (2007) Are exotic natural enemies an effective way of controlling invasive plants? *Trends in Ecology and Evolution* 22, 447–453.

Thomas, M.B. and Willis, A.J. (1998) Biocontrol – risky but necessary? *Trends in Ecology and Evolution* 13, 325–329.

Tisdell, C.A. (1990) Economic impact of biological control of weeds and insects. In: Mackauer, M., Ehler, L.E. and Roland, J. (eds) *Critical Issues in Biological Control*. Intercept, Andover, UK, pp. 301–316.

Trujillo, E.E. (2005) History and success of plant pathogens for biological control of introduced weeds in Hawaii. *Biological Control* 33, 113–122.

United States Environmental Protection Agency (2010) *Regulating Biopesticides*. Available at: http://www.epa.gov/opp00001/biopesticides/index.htm (accessed 29 April 2010).

van Beneden, S., Leenknegt, I., Franca, S.C. and Hofte, M. (2010) Improved control of lettuce drop caused by *Sclerotinia sclerotiorum* using Contans combined with lignin or a reduced fungicide application. *Crop Protection* 29, 168–174.

van Driesche, R.G. and Bellows, T.S. (1995) *Biological Control*. Chapman and Hall, New York.

van Driesche, R., Hoddle, M. and Center, T. (2008) *Control of Pests and Weeds by Natural Enemies: an Introduction to Biological Control*. Blackwell Publishing, Oxford, UK.

van Frankenhuyzen, K. (2009) Insecticidal activity of *Bacillus thuringiensis* crystal proteins. *Journal of Invertebrate Pathology* 101, 1–16.

van Lenteren, J.C. (2000) A greenhouse without pesticides. *Crop Protection* 19, 375–384.

van Lenteren, J.C., Babendreier, D., Bigler, F., Burgio, G., Hokkanen, H., Kuske, S., Loomans, A., Menzler-Hokkanen, I. Van Rijn, P., Thomas, M., Tommassini, M. and Zeng, Q.-Q. (2003) Environmental risk assessment of exotic natural enemies used in inundative biological control. *BioControl* 48, 3–38.

van Lenteren, J.C., Bale, J., Bigler, F., Hokkanen, H.M.T. and Loomans, A.J.M. (2006) Assessing risks of releasing exotic biological control agents of arthropod pests. *Annual Review of Entomology* 51, 609–634.

van Lenteren, J.C., Loomans, A.J.M., Babendreier, D. and Bigler, F. (2008) *Harmonia axyridis*: an environmental risk assessment for Northwest Europe. *BioControl* 53, 37–54.

Verma, M., Brar, S.K., Tyagi, R.D., Surampalli, R.Y. and Valero, J.R. (2007) Antagonistic fungi, *Trichoderma* spp.: panoply of biological control. *Biochemical Engineering Journal* 37, 1–20.

Vestergaard, S., Cherry, A., Keller, S. and Goettel, M. (2003) Safety of hyphomycete fungi as microbial control agents. In: Hokkanen, H.M.T. and Hajek, A.E. (eds) *Environmental Impacts of Microbial Insecticides*. Kluwer Academic Publishers, Dordrecht, the Netherlands, pp. 35–62.

Vinale, F., Sivasithamparam, K., Ghisalberti, E.L., Marra, R., Woo, S.L. and Lorito, M. (2008) *Trichoderma* – plant–pathogen interactions. *Soil Biology and Biochemistry* 40, 1–10.

Vlot, A.C., Dempsey, D'M.A. and Klessig, D.F. (2009) Salicylic acid, a multifaceted hormone to combat disease. *Annual Review of Phytopathology* 47, 177–206.

Waage, J.K. (1997) Biopesticides at the crossroads: IPM products or chemical clones? In: Evans, H.F. (chair) *Microbial Insecticides: Novelty or Necessity? BCPC Symposium Proceedings No. 68*. British Crop Protection Council, Farnham, UK, pp. 11–20.

Walters, D.R. and Fountaine, J.M. (2009) Practical application of induced resistance to plant diseases: an appraisal of effectiveness under field conditions. *Journal of Agricultural Science* 147, 523–535.

Walters, D., Newton, A. and Lyon, G. (eds) (2007) *Induced Resistance for Plant Defence: a Sustainable Approach to Crop Protection*. Blackwell Publishing, Oxford, UK.

Wang, S., Just, D.R. and Pinstrup-Andersen, P. (2006) Tarnishing silver bullets: Bt technology adoption, bounded rationality and the outbreak of secondary pest infestations in China. Selected paper prepared for presentation at the *American Agricultural Economics Association Annual Meeting*, Long Beach, California, 22–26 July 2006.

Weatherston, I. and Stewart, R. (2002) Regulatory issues in the commercial development of pheromones and other semiochemicals. *Use of Pheromones and Other Semiochemicals in Integrated Production. IOBC wprs Bulletin* 25, 1–10.

Whipps, J.M. (2001) Microbial interactions and biocontrol in the rhizosphere. *Journal of Experimental Botany* 52, 487–511.

Whipps, J.M. and Davies, K.G. (2000) Success in biological control of plant pathogens and nematodes by microorganisms. In: Gurr, G. and Wratten, S. (eds) *Biological Control: Measures of Success*. Kluwer Acdemic Publishers, Dordrecht, the Netherlands, pp. 231–269.

Whipps, J.M., Sreenivasaprasad, S., Muthumeenakshi, S., Rogers, C.W. and Challen, M.P. (2008) Use of *Coniothyrium minitans* as a biocontrol agent and some molecular aspects of sclerotial mycoparasitism. *European Journal of Plant Pathology* 121, 323–330.

Williams, T., Valle, J. and Vinuela, E. (2003) Is the naturally derived insecticide Spinosad compatible with insect natural enemies? *Biocontrol Science and Technology* 13, 459–475.

Witzgall, P., Stelinski, L., Gut, L. and Thomson, D. (2008) Codling moth management and chemical ecology. *Annual Review of Entomology* 53, 503–522.

Wraight, S.P. and Ramos, M.E. (2005) Synergistic interaction between *Beauveria bassiana* and *Bacillus thuringiensis tenebrionis* based biopesticides applied against field populations of Colorado potato beetle larvae. *Journal of Invertebrate Pathology* 90, 139–150.

Zimmermann, G. (2007a) Review on safety of the entomopathogenic fungi *Beauveria bassiana* and *Beauveria brongniartii*. *Biocontrol Science and Technology* 17, 553–596.

Zimmermann, G. (2007b) Review on safety of the entomopathogenic fungus *Metarhizium anisopliae*. *Biocontrol Science and Technology* 17, 879–920.

4 The Economics of Making the Switch in Technologies

So far in this book we have seen that IPM, to date, is used to varying degrees of intensity in nearly all annual protected and unprotected crop systems. However, this is far from a universal picture since not all growers have adopted this approach. Furthermore, for the majority of production systems, particularly in broad-acre agriculture, adoption and impact (as measured by reductions in use of conventional chemical pesticides) have been minimal. This lack of commercial adoption appears to run counter to various assertions by researchers of the appropriateness and even the superiority of these new techniques over the incumbent technology.

In the wider technology adoption context, this apparently irrational behaviour on the part of farmers has been the subject of a number of contributions within the economics literature.

Typically, the producer's decision of whether or not to adopt a novel technology can be conceived as a simple comparison of the relative net benefits derived from the novel and the incumbent technologies. The benefits are usually defined as utility, which for the commercial producer will relate primarily to profits. However, at least for the novel technology, farmers will have imperfect information on the costs of implementing and the benefits gained from the novel technology. Therefore, the farmer must make a judgement to form his expectation of the net benefits of using the new technology. Economists tend then to formalize the single technology adoption/replacement decision as one where adoption occurs if $E(U_n, z) > E(U_o, z)$, where n refers to the novel and o to the old technology and z is a range of farm characteristics and prices. This framework works well when we consider the adoption of single independent innovations, a good example of which could be the adoption of computerized record-keeping. However, there are a number of characteristics of the IPM adoption problem that are likely to considerably complicate our simple technology adoption decision criteria.

© CAB International 2010. *Biopesticides: Pest Management and Regulation*
(A. Bailey *et al.*)

Several possible reasons are cited. These include:

1. *Differential cost structures of pesticide and IPM technologies.* The significant research database and the widespread practical use of chemical pesticide technology ensure that what fixed costs are required are both mitigated by widely available information and diluted over many users. The fixed costs faced by users of this technology are a small fraction of the total costs of pest control. It has been argued by many authors (e.g. Cowen, 1991; Cowen and Gunby, 1996) that this situation is both an artefact of the technology (farmers need only limited information to ensure that control is effective, if not economical) and a product of past dynamic gains from adoption. In contrast, existing biocontrol and other alternative pest management technologies are very information and skill dependent. This factor ensures that potential adopters of this technology face significantly large fixed costs of adoption. However, these fixed costs are likely to diminish as these technologies gain more widespread use and, as such, dynamic gains are likely to be increasing albeit from a small base.
2. *Differential risk of technologies and risk preferences of producers.* The much more limited body of research data and practical experience with biocontrol and IPM adds to the degree of uncertainty surrounding the efficacy and economic viability of the techniques (e.g. Cowen and Gunby, 1996; Abadi Ghabim and Pannell, 1999; Pannell, 2003) and acts as a barrier to adoption. In addition, the uncertain effects of scaling in alternative technologies also hinder the adoption process (Griffiths *et al.*, 2008a). Furthermore, farmers' heterogeneous, and potentially averse, risk preferences are likely to result in a less than uniform, and probably suboptimal, adoption pattern of uncertain technologies (e.g. Antle, 1987; Yaron *et al.*, 1992; Cowen and Gunby, 1996). Potential risk aversion adds to the problem if expectations of the efficacy of alternative strategies based on biocontrol are skewed towards the 'downside' (Antle, 1987).
3. *Jointness caused by allocatable fixed factors* (Shumway *et al.*, 1984) (the sprayer machine). Jointness in the control of a range of pests, plant pathogens and weeds using conventional chemical control techniques, caused by use of common shared fixed assets, presents a further potential barrier to the adoption of alternative biocontrol techniques. The broader the range of biocontrol approaches integrated into alternative strategies the greater the potential gains to adopters. As such, a more integrated approach to biocontrol research, development and extension is likely to be important.
4. *Portfolio economies in IPM adoption.* While some substitution possibilities may exist between component techniques with IPM, there will probably exist some strong yield-enhancing or cost-reducing complementary relationships. These are often highlighted by ecologists working with biocontrol evaluation. Practical examples include flying biocontrol organisms that predate or act in the upper crop canopy and those ground-moving organisms that act lower down the plant being spatially additive. Likewise, biocontrol organisms that are active at different times of the cropped season can be considered as temporally additive. Therefore, IPM adaptations to systems and land-use practices which promote these different biocontrol agents will be

functional complements. However, where the presence of one IPM technique in a system results in improved performance in a second technique, where complementary techniques are super-additive, then there will be economies of adoption intensity, or portfolio economies. Griffiths *et al.* (2008b) have quantified some of these effects in the context of the rate of pest control provided by control agent assemblage complexity. If these intra-species effects scale up to the level of IPM component pest management technologies, then they will reinforce portfolio economies. Therefore, farmers must consider the potential benefits of adopting a specific IPM technique in the light of their currently adopted portfolio of technologies since $e_n = f(t_n, T)$, where e_n and t_n are respectively the efficacy and intensity of adoption of the nth technique and $(T = t_1, t_2, \ldots, t_{n-1})$ is the intensities of the set of currently used technologies or practices.

All of these factors will probably contribute towards an advantage of the incumbent technology. As Cowen and Gunby (1996) point out, in the competition between technologies which perform similar roles, history matters when one or both technologies possess increasing returns to scale in either production or use. For the user, the farmer, scale economies could arise from 'learning by doing', falling information costs of implementation and scale or size economies relating to subsequent and sequential application cost. In addition, external scale economies (and diseconomies) in the pest control process itself, associated with possibly reduced pest pressure (and diminished population of beneficial organisms) derived from the use of pesticides on neighbouring farms, can complicate the issue. If so, the incumbent technology, in this case chemical pest control, will probably benefit from positive feedback as scale economies increase the returns to both current users and new adopters. Even if the alternative, IPM, technologies also possess economies of scale, and really are superior to the incumbent, farmers will find it extremely costly to switch technologies in the medium term and potential gains will remain unrealized. Subsequent technology choice will probably then be 'path dependent' and chemical pest control will remain 'locked in' to the system until such a time when the incumbent technology fails, as resistance is encountered and new toxicity pathways are exhausted or society deems that the use of particular chemical pesticides is unacceptable.

All of these issues probably compound to make the adoption decision process cognitively complex. Furthermore, the issue of path dependency in both the incumbent and novel technologies probably promotes inertia.

A number of researchers have attempted to model the decision to adopt multiple, potentially interdependent, farming technologies. Examples in agriculture include Feder (1982), who considers the adoption of two technologies under uncertainty and where one technology has scale economies, and Dorfman (1996). These approaches consider the decisions of farmers faced with the options to not adopt or to adopt technology i, technology j or some combination of technologies $i, j, \ldots, S$.

Dorfman models the portfolio adoption process by considering each alternative adoption decision, for a two technology decision as either (0,0),

(0,1), (1,0) or (1,1), a problem with four potential 'technology bundles' or portfolios. Dorfman then applies a multinomial logistic regression model to explain adoption patterns of IPM and improved irrigation technologies in US apple farms. Clearly, the number of technology bundles to consider increases rapidly the more distinct technologies we consider. With three technologies the number of unique bundles increases to seven, with four technologies to 11, and so on. At this level of technology disaggregation, econometric identification of key parameters is near impossible since model likelihood functions are quite flat. Moreover, since the farmer must develop an expectation of the utility he could derive from each bundle, his cognitive burden is considerable. The farmer may find it quite easy to develop an expectation for bundles such as (0,0,0) or (0,1,0). However, any bundle including more than one technology will require the farmer to assess whether the expected utility from (1,1,0), never mind (1,1,1), is greater than or less than the sum of its parts. The potential for IPM portfolio economies to promote a positive path dependency resulting in the sequential adoption of IPM techniques could be strong if the technical 'incrementalism' referred to in Chapter 3 is significant.

Feder considers the decision facing a farmer with the option to adopt two interrelated technologies which differ in cost structure, one with constant average and marginal cost (Feder considers high-yielding crop varieties) and the other with a lump-sum start-up cost and therefore a declining average cost (the example used is irrigation tube wells). The technologies considered by Feder are interrelated in two ways. First, the adoption of both technologies together produces higher average yields than achieved when adopting only one of the technologies, a yield complementarity. Second, the adoption of the declining cost technology affects the perceived risk of the constant cost technology. In Feder's example, perceived risk of adopting higher-yielding crop varieties might be reduced by adopting irrigation technologies. It is then shown that an increase in the fixed cost of adopting irrigation, a smaller perceived risk complementary effect, a higher perceived risk of using improved crop varieties and increasing farm size (if farmers are risk averse as is often the case) can all reduce the adoption of higher-yield varieties. These results remind us that we must consider a broader definition of technological complementarity, than simple yield complementarity, in our assessment of IPM approaches and that scientific evaluation work on the interrelated risk effects of joint adoption could prove highly influential in the portfolio adoption decisions of farmers.

Faced with these types of adoption problems, it seems reasonable to assume that farms will probably retreat into a tactic of trialling, and subsequently adopting or dropping, individual technologies in a sequential manner. This approach may generate some beneficial portfolio approaches; however, it is likely that the choice of the first technology will condition the shape of the resulting technology portfolio. Equally, since we would expect that the expected utility derived from any technology, or set of technologies, would be conditional on farm characteristics, including geography, crop combinations and farmer skill sets, it would be quite difficult to make specific portfolio recommendations, although the science community could help here by identifying key technology complements for general situations or specific pest problems.

Pragmatic Promotion of Integrated Pest Management Adoption

Many authors who consider both the science and the economics of the IPM adoption process recognize the importance of the 'public good' aspect of alternative forms of pest control. This may be the case where the adoption of IPM reduces the potential external effects associated with pesticide use, such as effects on non-target organisms and the contamination of watercourses, whose costs are borne by society at large. Additionally, the adoption of IPM on one farm may enhance the benefits of IPM on neighbouring farms. The presence of external effects can justify calls for government action to equalize private incentives with the costs or benefits to society. Cowen and Gunby (1996) argue that 'path dependency' suggests that farmers are not necessarily irrational in choosing to continue to use an entrenched but suboptimal technology. However, society as a whole may be considered to be irrational if the public policy provision required to change to a socially more optimal path is not committed. Elsewhere in this book we consider the role of contracts in the food supply chain in this regard (see Chapter 7). While private contractual arrangements have been shown to be effective in inducing a switch away from pesticides to IPM in some production systems, this cannot be considered an option for systems which produce primarily for spot market sale. Here, government action remains the primary hope for breaking pesticide dependence. Agri-environmental policy (AEP) may then help to improve the financial return of IPM to farmers and close the technology gap opened up by scale economies in the incumbent chemical control technology.

Other government action may also help the process of adoption. As extensively discussed elsewhere in this book, the regulation of chemical pesticides provides an interesting case. The action of the regulator (the PSD, now CRD, as competent authority for Directive 91/414/EEC replaced by Regulation 1007/2009 in the UK case) may have the effect of slowing up the supply of newly approved compounds, by the application of stringent safety and testing requirements and the removal of existing compounds considered unsafe given new knowledge and standards. Much is written on these forces elsewhere in this book. Either mechanism could provide the demand pull that promotes an increased adoption of IPM technologies.

Thus, while there are many drivers acting to reduce the use of pesticides, and there are many alternative pest management strategies in development, adoption of these alternatives remains limited to date (particularly in the arable sector).

Lohr and Park (2002) considered the mix, or number, of insect management technologies adopted by organic farmers in the USA. Their data showed that over 30% of their sample of 1027 farms used more than five distinct technologies within their IPM portfolio. Lohr and Park subsequently used a count data model to consider how the number of technologies adopted, or the diversity of a farm's IPM portfolio, was influenced by various farm-specific characteristics. On the whole, highly educated and more experienced organic full-time farmers who operated on smaller farm areas producing a high proportion of horticultural crops adopted a wider range in insect pest-specific IPM technologies. IPM complexity, and by extension IPM success, then appears to hinge on scouting, information and an ability to understand complex systems. How IPM rolls out into more extensive crop systems, where human capital must be spread thinly, is of great interest. Public policy is likely to be important in the adoption of IPM in broadacre agriculture.

Agricultural and agri-environmental policy

AEP can, in principle, give rise to landscapes that can support a large number of arthropods such as pests and their enemies and can help to overcome some of the problems of 'path dependence' described by Cowen and Gunby (1996). Indeed, the types of habitat manipulation available via the AEP should, in principle, give rise to the provision of many natural predators of value to cereal producers. Holland and Oakley (2007) even identify those manipulations most likely to yield the preferred outcomes (i.e. hedgerows, field margin strips, beetle banks). They also note that within-field activities such as non-inversion tillage need to be considered given the growing implementation of this method in practice.

Currently, there are a number of strong drivers in the agricultural policy domain to suggest that farmers will adopt a number of different strategies, both consciously and unconsciously, towards pest management. In the UK and the EU more generally agricultural policy has undergone significant modifications in recent years. These changes have resulted in farmers being targeted by two strands of financial aid or incentives. First, the reform of the Common Agricultural Policy has seen the introduction of the Single Farm Payment Scheme and Cross Compliance. The Single Farm Payment is a payment made to farmers which is decoupled (partially in many cases) from current production. The Single Farm Payment is meant to replace all previous forms of commodity support so that the EU can move towards compliance with the demands placed on agricultural policy as a result of the General Agreement on Tariffs and Trade/World Trade Organization trade negotiations. The UK Department for Environment, Food and Rural Affairs (DEFRA, 2007) reports that the Single Farm Payment attracted £2.4 billion in expenditure in 2006. This compares with £346 million for AEP.

Initially, the Single Farm Payment is based on an agreed benchmark anchored to each farm's historic 'policy' receipts; however, a phased shift towards a flat-rate payment system, differentiated only by region, is under way. In order to qualify for receipt of the Single Farm Payment, farmers need to satisfy the Cross Compliance standards and requirements (DEFRA, 2004) of which there are two types.

1. *Good Agricultural and Environmental Conditions (GAEC)*: maintain land in appropriate condition – includes requirements such as protection of hedges, stone walls and watercourses. Defined by each member state in order to protect soils and ensure minimum level of maintenance and avoid deterioration of habitats (especially for non-cultivated land, i.e. pasture) (Kirkham *et al.*, 2004).
2. *Statutory Management Requirements (SMRs)*: secure compliance with various EU Regulations and Directives – 18 in total.

Second, there has been a significant reform and reconfiguration of AEP. From early 2005 the UK introduced the Environmental Stewardship scheme:

> ES is a new agri-environment scheme which provides funding to farmers and other land managers in England who deliver effective environmental management on their land. The scheme is intended to build on the recognized success of the ESA and CSS and its primary objectives are to:
>
> 1. Conserve wildlife (biodiversity)
> 2. Maintain and enhance landscape quality and character
> 3. Protect the historic environment and natural resources
> 4. Promote public access and understanding of the countryside
>
> Within the primary objectives, it also has the secondary objectives of: Genetic conservation; and Flood management.
>
> (DEFRA, 2005: 2)

Environmental Stewardship is composed of the Entry Level Stewardship scheme (ELS), the Organic Entry Level Stewardship scheme (OELS) and the Higher Level Stewardship scheme (HLS). These schemes replace existing AEP such as Environmentally Sensitive Areas scheme (ESA), the Countryside Stewardship Scheme (CSS), the Organic Farming Scheme (OFS) and the Wildlife Enhancement Scheme (WES).

In terms of pest management activities, both the ELS and OELS are highly relevant. Beginning with the ELS, this scheme is open to all land managers in England with entry guaranteed providing scheme requirements are met. This requires the land manager to select a number of environmental commitments which earn points that are explicitly stated as part of the scheme and, once the threshold of 30 points/ha is attained, entry is assured. Entry is currently for 10 years with a contract initially for 5 years, which can then be extended. Management options include things such as protection and management of boundary features, woodland edges, the conservation of soils and historic and landscape features. The ELS is designed to be simple with little need for expert knowledge in terms of the form to be completed other than that held by a competent farmer. The intention of the ELS is to secure basic-level environmental management over as large a territory as possible. Currently, the financial payment is set at £30/ha per annum. The OELS is very similar in terms of how it is intended to operate albeit with slightly modified objectives and management options. The OELS attracts slightly higher payments at £60/ha per annum.

Currently, ELS and OELS farm management requirement go beyond Cross Compliance. However, the management options selected will complement, but not remove, obligations arising from Cross Compliance.

Concurrent with these changes to agricultural policy there have also been efforts in the UK to minimize the negative externalities associated with pesticide use. Specifically, there has been an industry initiative/programme called the Voluntary Initiative on pesticides. The Voluntary Initiative has attempted to bring about best practice in pesticide use by initiating research, training and communication, and stewardship (www.voluntaryinitiative.org.uk). The Voluntary Initiative was introduced in April 2001 after a long debate about the appropriate policy mechanism to employ to deal with pesticide externalities. The Voluntary Initiative was initially meant to last 5 years but it has recently been extended on a 2-year rolling basis. However, there are still strong reasons to assume that a pesticide tax will be considered again, given the Voluntary Initiative's 2-year rolling window of continued operation.

The Voluntary Initiative, as its name suggests, consists of a set of voluntary industry actions with the aim to reduce the impact of pesticides on the environment. Examples of the types of programmes in the Voluntary Initiative include research, training, communication and stewardship. The package of measures was devised by the Crop Protection Association in collaboration with various industry bodies such as the National Farmers Union. It has been estimated that the cost of the programme over the 6 years it has been in operation is approximately £45 million (Voluntary Initiative Steering Group, 2006).

The significance of the Voluntary Initiative can be seen in the impact it has had on the development of land options as part of the ELS. In particular, the Voluntary Initiative supported a research project demonstrating that skylark plots, which attract points for the ELS, can significantly increase skylark fledgling survival. Also the Voluntary Initiative introduced Crop Protection Management Plans (CPMPs). These plans are a self-audit of farm-level crop protection activities. The CPMPs were included in the ELS and at present it is estimated that some 1.5 million ha are covered by approved plans. Controversially, however, DEFRA have now removed this option from the scheme.

The inclusion of the CPMP in the ELS provided a farmer with 2 points/ha. The CPMP requirements were reasonably general, covering aspects of pesticide use such as storage, handling and application. There was also an emphasis on the choice of pesticide product as well as the adoption of non-chemical pest management options. Those practices considered here included the use of deliberate crop rotations, cultivations practice and timing, the planting of resistant crop varieties and the use of natural predators via habitat enhancements such as beetle banks and field margins.

In an effort to understand what farmers are currently doing with respect to pest management we can consider what types of actions and activities have been adopted as part of the ELS. Boatman *et al.* (2007) have examined the uptake of the ELS to date and their results provide an interesting insight into how this scheme has influenced pest management strategies on farms.

Currently some 27,000 farmers participate in the various Environmental Stewardship schemes compared with some 170,000 non-participants. As we might expect, farmers adopt management and land-use practices that make least change to current farming practices. Boatman *et al.* (2007) estimate that 60% of ELS activities entered into the scheme were already being done in the desired manner, which raises questions regarding over-compensation. However, there has been a large increase in the land area and number of agreements under AEP since 2005 and the introduction of the Environmental Stewardship scheme.

In terms of land area the ELS agreements cover some 3.5 million ha with the highest proportion being in eastern regions. Cereal farmers are the largest group of participants by type, both in terms of number and area. It is also the case that cereal/cropping farms have adopted the largest number of options in the ELS per farm: on average, 8.6 options for cereals and 9.0 for general cropping. If a farm has adopted a large number of practices it is possible that they have adopted some of the options that proved less popular overall. The reason given by farmers for the adoption of particular options was the points gained. This may have been particularly acute in the case of arable farms since, for these farm types, gaining points was difficult because of the existing farming systems, landscape and land-use patterns.

So what practices and options have proved popular? Again as might be imagined, those options which require little in the way of systems changes, little significant land-use change or those which alter land use on marginally productive areas have proved popular. These include hedge and ditch management, field corner management on arable farms (popular in the east), and 4 m and 6 m buffer strips on cultivated land.

Those options which have proved far less popular can be summarized as those which require additional effort, significant investment or land-use change on more productive land. These include the use of buffer strips on intensive grassland, wild bird seed mix or pollen and nectar mix on set-aside, beetle banks, skylark plots, conservation headlands and uncropped cultivated margins on arable land, and all options to encourage a broader range of crop types on farms.

As we have already noted, the CPMP from the Voluntary Initiative had been included as a management option within the ELS. From the most recent data available we can see that this option was adopted by 10,164 farms or 39.5% of all farms in the ELS. As might be expected, many of the farms adopting CPMPs are cropping dominated farms (see Table 1.16 in Boatman *et al.*, 2007: 28), some 69.3% of cereals farms and 76.1% for general cropping farms. The average area entered into CPMPs is 195.1 ha. However, we cannot be sure as to whether the ELS has helped to pump-prime the Voluntary Initiative or whether the ELS essentially over-compensates existing activities here.

We can also consider the extent to which financial incentives have encouraged farmers to adopt non-pesticide disease and pest management practices. From the data we can see that the adoption of practices that may provide non-chemical pest control is relatively low. Some important incentives built into the ELS are summarized in Table 4.1 and here we note the relatively poor take-up of those options of potential benefit to conservation biocontrol, namely beetle banks, pollen and nectar mix on set-aside and conservation headlands.

Table 4.1. Financial incentives to encourage pest management in the Entry Level Stewardship scheme. (From DEFRA, 2005; Boatman et al., 2007.)

Option	Cereal farms (%)	Cropping farms (%)	Points attained
EF2 (Wild bird seed mix)	17.7	21.0	450[b]
EF3 (Wild bird seed mix on set-aside)	2.6	3.25	85[b]
EF4 (Pollen and nectar mix)	11.3	13.1	450[b]
EF5 (Pollen and nectar mix on set-aside)	0.9	0.9	85[b]
EF7 (Beetle banks)[a]	2.2	2.6	580[b]
EF8 (Skylark plots)	3.6	3.4	5[c]
EF9 (Conservation headlands in cereals)	0.4	0.7	100[b]

[a]This equates on average to 0.2 ha or 1 km × 2 m wide.
[b]Points per hectare.
[c]Points per plot.

If we turn to the OELS, 167,000 ha were entered into the scheme mostly located in the south-west. The area of uptake is greatest by lowland grazing livestock farms and mixed and dairy farms. The percentage of cereals and general cropping farms entering the scheme is very low; although cereal farmers have enrolled the largest total area, the proportion of all cereal farmers is much lower than for dairy or mixed. Even so, it is noteworthy that very few organic farmers have adopted either IPM beneficial beetle banks or skylark plots, which might surprise some readers.

Research supported by RELU grant RES-224-25-0093, using a similar approach as used by Lohr and Park (2002), presents an interesting picture of current (2007) IPM adoption in UK arable agriculture. Data include observation on 547 farms. While the data set used in this case includes conventional farms (92.6%), organic (5.4%) and farms with both conventional and organic activity (1.9%), a high proportion of the sample does report adoption of a relatively large number of IPM technologies on their farms. Respondents were asked to report their use and interest in a prescribed set of IPM technologies. For each technology considered, respondents were asked to indicate whether they currently use, have discontinued using, will consider using in the near future or will not use each technology. The results of this question for the 17 pesticide alternative practices are reported in Fig. 4.1.

The results reported in Fig. 4.1 show a strong gradient of differential adoption among the pest management practices considered, with some quite widely adopted and others that are far less prevalent. Many of these results are unsurprising a priori. An interesting (if less than surprising) result, however, is the relatively large number of farmers using improvements in field margins. This type of field management requirement is part of many of the AEP contracts and requires only a marginal addition to a land management practice to that required by the Single Farm Payment scheme.

It is interesting to note that very few of the technologies appear to have been discontinued following a trial phase. The highest level of discontinuation of a technology is 15% for both pheromone monitoring and disease- or

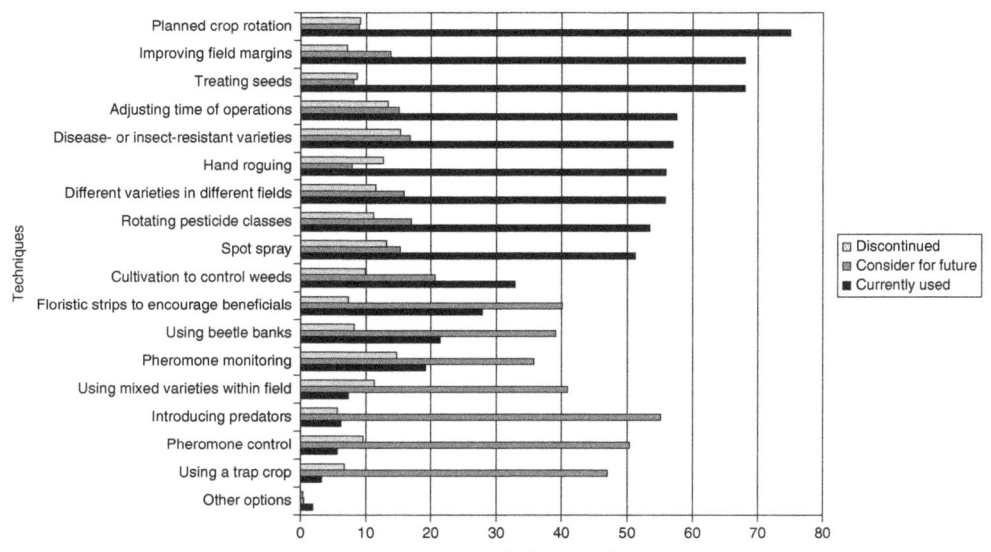

Fig. 4.1. Adoption of pest control methods. (From Bailey et al., 2009.)

insect-resistant varieties. This would suggest that, if farmers do trial a practice, then they are highly likely to continue to use, or to fully adopt, that practice. Of the technologies being considered there would also appear to be a degree of resistance to even trying some of them, in particular planned rotations, treating seeds and hand roguing, although there appears to be a reasonable likelihood that some of the currently less popular technologies might be considered for adoption at some point in the future.

The data report that some 61% of farmers are using more than five IPM technologies within their pest management portfolios. Many of these technologies were probably adopted by farmers in order to comply with AEP contracts. However, these data do suggest that broad-acre farmers are employing diverse and complex IPM portfolios and that reliance on chemical methods alone is rare. More specifically, just 6% of these farmers reported that they currently use introductions, within which we can consider biopesticides an option. However, a further 51% of the sample expressed an interest to trial these technologies in the near future. More generally, those technologies which can be considered in some way 'embodied' in capital or material goods are considered positively by 55% of the sample while 63% of the respondent farmers consider the more 'disembodied' techniques positively. While arable farmers appear to favour IPM interventions which they can provide from their own resources of land and labour, they do appear willing to purchase IPM products.

The adoption data collected in this project were used to uncover the different types of IPM portfolios farmers use in the wider pest management context. A principal components analysis revealed that the 17 pest management techniques could be summarized by four distinct portfolio approaches. This approach was used by Rauniyar and Goode (1992) when they considered the adoption of seven technologies by maize farmers in Swaziland. These portfolios are described in Table 4.2.

It would appear that the specific problems faced by farmers in their crop systems have a significant bearing on portfolio choice and the portfolio names suggest this. Regression analysis was then used to investigate further

Table 4.2. Distinct integrated pest management portfolios in UK arable crops. (From Bailey et al., 2009.)

Portfolio 1: 'Intra-crop biocontrollers'	Portfolio 2: 'Chemical "users"/conservers'	Portfolio 3: 'Extra-crop conservation biocontrollers'	Portfolio 4: 'Weed-focused farmers'
Trap crops	Pheromones	Field margins	Cultivate weeds
Mixed varieties	Different varieties	Floral strips	Crop rotation
Introductions	Resistant varieties	Beetle bank	Timing of operations
Pheromones	Spot spraying		Hand roguing
Different varieties	Treated seeds		
	Rotate pesticide classes		

determinants of portfolio choice by farms. This analysis sheds very little light on the 'Intra-crop biocontrollers'. However, there appeared to be some significant relationships in the other three cases. Factor scores for the second portfolio, 'Chemical "users"/conservers', appeared to be associated with increasing cropped areas (larger arable operations), with higher frequency of insecticide application, and involvement in the ESA and membership of the Voluntary Initiative. Organic status, perhaps not surprisingly, was negatively related to this portfolio approach.

For 'Extra-crop conservation biocontrollers' there appeared to be a statistically significant negative relationship with the number of insecticidal applications per crop; on the other hand, a positive statistically significant relationship with the proportion of land with tenant rights and membership of the Voluntary Initiative. Certainly, the absence of tenant rights would probably form a barrier to the adoption of habitat manipulations that require some significant investment, beetle banks for example. There were four statistically significant relationships between farm characteristics and scores of 'Weed-focused farmers'. Here, livestock farms with high levels of tenant rights and those engaged in the HLS were less likely to opt for Portfolio 4, while organic farms appeared to be more likely to adopt Portfolio 4.

One further regression model employed these data, alongside a range of farm characteristics, to assess the impact of portfolio choice on differential rates of insecticide application intensity (number of insecticide applications per crop) across farms.

Results suggested that arable farmers who rely on independent crop consultants for their spray advice, are members of the ELS and have adopted Portfolio 2, 'Chemical "users"/conservers', tended to spray for insect pests more frequently. The results also suggested that farmers who follow practices described by IPM Portfolio 1, the 'Intra-crop biocontrollers', appeared to have applied chemical insecticides in a less intensive manner than did their peers. Two further 'nearly' significant relationships were those for membership of the Voluntary Initiative and the proportion of land farmed conventionally. Both of these results tentatively suggested a positive effect on applications, albeit at a confidence level of only 89.3% and 87.0%, respectively. The two statistically significant relationships for Portfolios 1 and 2 confirmed prior expectation in terms of sign since it seems reasonable to assume that intra-crop biocontrol strategies should reduce reliance on chemical pesticides while practices aimed at reducing selection pressure on pesticides might still result in a significant reliance on pesticides. Furthermore, while it is not surprising that efforts to control weed problems have little impact on insecticide use, as found for Portfolio 4, it is surprising, if not disappointing, to find no evidence of an impact of using Portfolio 3, 'Intra-crop conservation biocontrol' practices, on insecticide use. Moreover, it is interesting to note that this research found that membership of the ELS, with its focus on environmental land-use change and CPMPs, was counter-intuitively correlated with a greater intensity of insecticide applications. Likewise, while statistical power was lacking, the positive effect of the Voluntary Initiative on insecticide spray intensity was also striking. Such results call for further work to uncover in more detail the impact of the Voluntary Initiative on pesticide use, given the UK Government's current reliance on this policy for pesticide use and impact reduction.

Recently the potential scaling impacts of IPM and biodiversity have been considered by the scientific (Griffiths *et al.*, 2008a) and policy-making (Franks and McGloin, 2007) communities. The potential for farmers to create, at least local, network external benefits in the provision of biocontrol and other conservation goals is now being considered. To this end, coordinated or cooperative bids submitted by groups of neighbouring farmers for collective AEP funding could provide the key to gaining otherwise elusive scale and network benefits from agro-ecosystem services.

The picture we see of a rather passive adoption of IPM portfolio practice, induced in no small measure by AEP, is likely to require farmers to employ extensive crop-walking if benefits from adoption are to be gained. Since farmers are unlikely to have chosen ELS options in a conscious effort to limit their reliance on chemical pesticide, we cannot expect farmers to subsequently modify their pest control strategies unless they already spray reactively in response to economic damage thresholds. While economic damage thresholds are relatively simple to understand and apply when considering chemical control alone, the cognitive process of assessing the benefits from spraying becomes excessive in an IPM context. The identification of the presence of a damage agent is not enough. The farmer needs to consider what other practices or pest control mechanisms are 'present' in a given site, how quickly and effectively they might be expected to act, and the risk of their failure, before an intervention, which could be chemical pesticide, a biopesticide or a pheromone, can be decided upon. Such complexity of decision calls for a decision support tool – preferably one that includes economic price information alongside pest control response functions. Our current state of IPM knowledge is some way off the level needed to produce such a set of rules.

Key Challenges for Further Research and Development in Integrated Pest Management

Research and development activity is typically split into separate activities. Traditionally, the public sector funds fundamental research that discovers new technologies through the acquisition of new knowledge. It is widely believed that the public sector should do this since private individuals and firms will not fill this role, although examples of fundamental science in the private sector do exist. The product of this fundamental science is then left to the private sector to develop in the form of patentable products which embody the new technology. Patent holders can then exclude other producers from producing, and non-purchasers from using, their technology while recovering their developmental costs through the price of the product.

There exists a split, then, between the research and developmental phases of technology innovation and delivery, which potentially suits patentable new technologies but leaves others forms of technology behind. Further, if farmers adopt portfolios rather than individual IPM components separately, then the commercial exploitation of IPM could easily suffer.

Either private innovators need to recognize that the success of their product-based technology requires the parallel development of system-based portfolio partner technologies and fund development of both, or government needs to take up the challenge of funding the development of systems, or disembodied, IPM components.

Future research is required in two areas.

1. Further research to understand the role different technologies play within alternative IPM portfolios in terms of both yield and cost and risk complementarity. We need to understand the type of relationship between IPM components such that we can prescribe portfolio mixes of techniques which build both improved efficacy and sufficient resilience into portfolios.

2. Research and evaluation must be conducted on an extensive scale. This is unlikely to be affordable if conducted by the public sector and especially if conducted using conventional experimental control approaches. This work must also consider the dynamic time frame of problem identification, action and outcome of IPM portfolios. As Griffiths *et al.* (2008a) point out, we do not know the effect of scaling up individual IPM component technologies, let alone know whether IPM portfolios exhibit increasing, constant or decreasing returns to scale. Given obvious budget constraints and the need to consider the performance of diverse technology sets in complex environmental settings, the logical approach will probably follow the approach used in the farm-scale evaluation of herbicide-tolerant transgenic crops in the UK (e.g. Firbank *et al.*, 2003; Perry *et al.*, 2003).

Both of these strands of research are required in order to:

1. Redesign AEP policy so as to promote a more appropriate set of IPM portfolios that do reduce the commercial use of chemical pesticides in farming systems.
2. Consider the reintroduction of CPMPs redesigned to explicitly promote IPM.
3. Persuade farmers of the efficacy, resilience and compatibility of IPM portfolios in their farming systems and encourage them to rely upon the action of the portfolio mix before reaching for chemical remediation.
4. Generate IPM portfolio specific decision support protocols, allowing farmers to select the practices available at, and other key features of, each site, for crop protection interventions.

References

Abadi Ghabim, A.K. and Pannell, D.J. (1999) A conceptual framework of adoption of an agricultural innovation. *Agricultural Economics* 21, 145–154.
Antle, J.M. (1987) Econometric estimation of producers' risk attitudes. *American Journal of Agricultural Economics* 69, 509–522.

Bailey, A.S., Bertaglia, M., Fraser, I.M., Sharma, A. and Douarin, E. (2009) Integrated pest management portfolios in UK arable farming: results of a farmer survey. *Pest Management Science* 65, 1030–1039.

Boatman, N., Jones, N., Garthwaite, D., Bishop, J., Pietravalle, S., Harrington, P. and Parry, H. (2007) *Evaluation of the Operation of the Environmental Stewardship, DEFRA Project Number MA01028, Final Report*. Central Science Laboratory, York, UK.

Cowen, R. (1991) Tortoises and hares: choice among technologies of unknown merit. *The Economic Journal* 101, 801–814.

Cowen, R. and Gunby, P. (1996) Sprayed to death: path dependence, lock-in and pest-control strategies. *The Economic Journal* 106, 521–542.

DEFRA (2004) *Single Payment Scheme. Cross Compliance Handbook for England*, 2005 edn. Department for the Environment, Food and Rural Affairs, London.

DEFRA (2005) *Environmental Stewardship: Look After Your Land and Be Rewarded*. Rural Development Service, Department for the Environment, Food and Rural Affairs, London.

DEFRA (2007) *Agriculture in the United Kingdom 2006*. Department for the Environment, Food and Rural Affairs, London.

Dorfman, J. (1996) Modelling multiple adoption decisions in a joint framework. *American Journal of Agricultural Economics* 78, 547–557.

Feder, G. (1982) Adoption of interrelated agricultural innovations: complementarity and the impact of risk, scale, and credit. *American Journal of Agricultural Economics* 64, 94–101.

Firbank, L.G., Heard, M.S., Woiwod, I.P., Hawes, C., Haughton, A.J., Champion, G.T., Scott, R.J., Hill, M.O., Dewar, A.M., Squire, G.R., May, M.J., Brooks, D.R., Bohan, D.A., Daniels, R.E., Osborne, J.L., Roy, D.B., Black, H.I.J., Rothery, P. and Perry, J.N. (2003) An introduction to the farm-scale evaluations of genetically modified herbicide-tolerant crops. *Journal of Applied Ecology* 40, 2–16.

Franks, J.R. and McGloin, A. (2007) Joint submissions, output related payments and environmental cooperatives: can the Dutch experience innovate UK agri-environmental policy? *Journal of Environmental Planning and Management* 50, 233–256.

Griffiths, G.J.K., Holland, J.M., Bailey, A. and Thomas, M.B. (2008a) Efficacy and economics of shelter habitats for conservation biological control. *Biological Control* 45, 200–209.

Griffiths, G.J.K., Wilby, A., Crawley, M.J. and Thomas, M.B. (2008b) Density-dependent effects of predator species-richness in diversity–function studies. *Ecology* 89, 2986–2993.

Holland, J. and Oakley, J. (2007) *Importance of Arthropod Pests and Their Natural Enemies in Relation to Recent Farming Practice Changes in the UK. Research Review No. 64*. HGCA, London.

Kirkham, F.W., Gardner, S.M., Critchley, C.N.R. and Mole, A. (2004) Using GAEC cross-compliance to protect semi-natural habitats on farmland and unused agricultural land. A discussion paper prepared for LUPG. http://www.lupg.org.uk/pdf/pubs_gaec_habitats_final_lupg[1].pdf (accessed April 2009).

Lohr, L. and Park, T.A. (2002) Choice of insect management portfolios by organic farmers: lessons and comparative analysis. *Ecological Economics* 43, 87–99.

Pannell, D.J. (2003) Uncertainty and adoption of sustainable farming systems. In: Babcock, B.A., Fraser, R.W. and Lekakis, J.N. (eds) *Risk Management and the Environment: Agriculture in Perspective*. Kluwer Academic Publishers, Dordrecht, the Netherlands, pp. 67–81.

Perry, J.N., Rothery, P., Clark, S.J., Heard, M.S. and Hawes, C. (2003) Design, analysis and statistical power of the farm-scale evaluations of genetically modified herbicide-tolerant crops. *Journal of Applied Ecology* 40, 17–31.

Rauniyar, G.P. and Goode, F.M. (1992) Technology adoption on small farms. *World Development* 20, 275–282.

Shumway, C.R., Pope, R. and Nash, E. (1984) Allocatable fixed inputs and jointness in agricultural production: implications for economic modeling. *American Journal of Agricultural Economics* 66, 72–78.

Voluntary Initiative Steering Group (2006) The Future for the Voluntary Initiative. Proposals from The Voluntary Initiative Steering Group Submitted to Lord Rooker DEFRA Minister of State for Sustainable Farming and Food. Available at: http://www.voluntaryinitiative.org.uk/_Attachments/resources/1084_S4.pdf (accessed April 2009).

Yaron, D., Dinar, A. and Voet, H. (1992) Adoption of innovation by family farms in the Nazareth region in Israel. *American Journal of Agricultural Economics* 74, 361–370.

5 The Regulation of Biopesticides: an International Analysis

The regulation of biopesticides takes place within a regulatory system that was designed for the regulation of synthetic (chemical) pesticides. Hence, the system has a number of features that did not make it amenable to the registration of biological pesticides and a number of adjustments have had to be made to the system to facilitate their registration. In this chapter there is a discussion of the Biopesticides Scheme in the UK and the equivalent Genoeg Scheme in the Netherlands. The chapter concludes by drawing on the lessons learned from experience in a range of countries to set out design principles for a better regulatory system for biopesticides.

Within the EU, the registration of pesticides involves both member states and the European Commission. In the USA, regulation is principally undertaken at the federal level by the EPA but there is some involvement of the individual states. The OECD and the World Health Organization (WHO) also have an influence on pesticides regulation and their role is discussed more fully later in the chapter.

Limitations of the Current Regulatory Model

The system of pesticides regulation was established to deal with chemical pesticides, which remain the predominant type of plant protection product. As Waage notes (1997: 11), 'Biopesticides, being living organisms, have properties which make their design, production and use potentially very different from that of chemical pesticides. However, present and planned development, production and delivery of biopesticides usually follows a chemical pesticide "model".' Thus, 'The pesticide regulation system has been developed for regulating conventional chemical sprays developed by large companies and several aspects of it do seem to act as barriers to the further development and commercialization of alternative control methods'

(ACP, 2004: 34). One of the objections made by those opposed to pesticides is 'how pesticides are spread directly into nature, unlike industrial chemicals' (Blok et al., 2006: 321). Given that biopesticides come from nature, these objections do not apply in the same way, but this has not been fully recognized by the regulatory system, which has taken insufficient account of the specific benefits of biopesticides.

Waage argued (1997: 16): '[It] is not the industry alone, but the entire pesticide regulatory process which has not adapted itself to the new opportunities which biopesticides provide.' Biopesticide development found itself 'locked into an inflexible and unimaginative chemical pesticide model' (Waage, 1997: 14). In the years since Waage's article was written, considerable progress has been made towards the reduction of the regulatory barriers to biopesticide availability, not least through initiatives such as the UK Biopesticide Scheme discussed later in this chapter.

Pesticide Registration in the UK

The UK operates within an EU regulatory system for pesticides that, as one would expect in a system based on multi-level governance, is made up of a two-tier system of registration. The regulatory regime is currently in a transitional period where some major pieces of national legislation are being replaced by new, harmonizing EU legislation, a process that is discussed more fully later in the chapter. Pesticides are composed of specific compounds designed to adversely affect a pest and are combined with inert substances to improve their handling, application and effectiveness. Active ingredients are assessed at a Community level for inclusion on a positive list (known as Annex 1). Products containing chemicals listed on Annex 1 must then be assessed and registered by member states. Active ingredients in existing plant protection products are reviewed as part of a rolling programme that is focused on products that were already on the market before the mid-1990s. The maximum permitted residue levels for pesticides in crops (MRLs) are laid out in legislation. These are not safety levels, although they may be perceived as such by consumers, but are based on the maximum residue that should be present if a pesticide has been used according to its approval.

Within the UK, what was the PSD but became in April 2009 the CRD, constituted as a directorate of the Health and Safety Executive (HSE), has been the agency responsible for the registration of agricultural pesticides. In response to the Hampton Review of the regulatory sector, PSD became a distinct entity within the HSE in April 2008. 'The merger is based on PSD maintaining its identity, structure and current business' (PSD, 2008a: 3). Before April 2009 other pesticides known as biocides were dealt with by the HSE from an office in Bootle. Biocides are now also dealt with by the CRD, still from the Bootle office, while what was PSD continues to operate from York. How these geographical arrangements will develop in the longer term remains to be seen, but as a bureaucratic joke what was PSD is sometimes referred to as 'Bootle East'. In this chapter, we shall continue to refer to PSD in relation to events before April 2009, but use CRD to refer to current practice.

The basic framework for PSD's (and CRD's) activities is set out in the Food and Environment Protection Act 1985 and the Control of Pesticides Regulations 1986. The Plant Protection Regulations 1995 implement the European Directive (91/414/EEC), which regulated plant protection products. 'Within the EU's mutual-recognition framework for regulatory approval and licensing, it sought to be the market leader and took an active part in pesticides standard-setting' (Hood et al., 2003: 126). PSD was originally part of the Ministry of Agriculture, Fisheries and Food (MAFF), but was constituted as a quasi-independent executive agency in 1993. After the dissolution of MAFF, it subsequently reported to the Department for Environment, Food and Rural Affairs (DEFRA) where it also advises ministers on pesticide issues. It had a staff of 178 in 2008, 115 of whom are scientists.

CRD's policy work is funded by a grant from DEFRA. Its approvals and registration work is undertaken on a cost recovery basis, through charges to the firms seeking registration and a levy on approval holders. The levy is designed to cover the cost of reviewing older pesticides and monitoring pesticides use and residues in food. It is based on a percentage of annual turnover arising from sales of approved pesticides. In 2006/7 out of a PSD total income of just over £13m, £5.3m came from the policy grant, £3.8m from the levy and £3.2m for fees, the balance being made up by incidental income.

Before granting marketing approval to a product conferring 'plant protection', the CRD evaluates data on human toxicity, environmental toxicity, pest control efficiency and potential to leave a residue on treated food crops. Recommendations for approval by the CRD are then forwarded to the ACP, a committee of independent experts. PSD and HSE provided the secretariat for the ACP but 'we are separate from, and independent of, these organisations' (ACP, 2005: 21). It would not be practical for the ACP to deal with every minor amendment to an approval so the Committee deals directly with the most important decisions, e.g. first commercial level approval of active ingredients new to the UK or any applications which appear to raise new concerns about safety. The final approval decision is taken by the responsible minister acting on the advice of a civil servant, but the minister is unlikely to overturn a recommendation made by the ACP. 'Advice from the ACP comes forward in written form from the Secretariat of the ACP to officials in Departments ... Once received, the advice is acted on directly by officials on behalf of Ministers – this is the case for technical issues and routine regulatory decisions' (RCEP, 2005: 79).

The Royal Commission on Environmental Pollution (RCEP) report on *Crop Spraying and the Health of Residents and Bystanders* was critical of the ACP, arguing in particular that its 'present approach may be conservative and protective in its treatment of targets' (RCEP, 2005: 59). The RCEP report drew a robust response from the ACP which sought to 'draw attention to various errors of fact and logic in their report. We note that several of its important conclusions appear to have been reached after what we consider to be an incomplete consideration of the relevant evidence' (ACP, 2005: 4). In its response, the RCEP claimed that 'We do not set out to criticise the ACP or suggest that UK practice is in any way less rigorous than elsewhere' (RCEP, 2006: 5). Nevertheless, 'we remain concerned that the ACP seems unable or unwilling to accept that most of its advice to Ministers is based on an implicit judgement, in a context of scientific uncertainty, about the relative importance of public concerns about human health and well being' (RCEP, 2006: 7–8).

Minor use crops: commercially significant but lacking protection

The term 'minor uses' is in many ways an unfortunate one as it encompasses field vegetable crops and fruit that are of some significance, such as carrots and peas. A minor use is defined in terms of one that occupies less than 50,000 ha so that potatoes and sugarbeet are above this dividing line in the UK. Brassicas are below this line, but, on the whole, there are very few crops in the 15,000–50,000 ha bracket. What is characteristic of minor uses is that the minor crops typically have available to them, given the relatively small size of the market, a small number of plant protection products that do not cover all the pest, disease and weed problems of that crop, or fail to provide choice between products to minimize the development of resistance to single pesticides. The loss of one active substance or its availability for use on minor crops could cause major difficulties and ultimately a situation could arise in which the crop could not be grown economically.

Carrots and parsnips provide an illustration of the problems that can arise. For carrot fly control, there is a now a diminishing range of active ingredients available that can deal with the problem. The Commission proposals under the revision of Directive 91/414/EEC could remove further substances which are important for the protection of carrots and parsnips with a potential for the loss of all yield. '[The] majority of the currently approved herbicides would no longer be available, and the control of some major weed species (e.g. mayweeds) would not be possible.' This could have significant commercial consequences. 'Weeds affect quality in terms of size, grade and uniformity. This is particularly important for baby carrots for quick-freezing or canning and fresh market. Failure to meet specifications can result in crop rejections or no sales' (PSD, 2008b: 36).

In 2003 PSD set up the Minor Uses Network to provide information and advise it on issues affecting the minor use sector. The coverage of the group was actually wider than 'minor' crops to ensure that niche pest problems on major crops are adequately covered. This Network was not operating in 2009, but there was discussion about resurrecting it. The existence of the UK Minor Uses Network stimulated the Commission to set up the Minor Users Steering Group. This represented all member states, but in practice there were a relatively small number of member states who are very active. Like the UK Minor Uses Network, a primary function was information exchange. A participant commented in interview: 'Looks at common problems and solutions, gets companies to develop actives on [a] broader range of crops and situations ... Helped get more authorisations, enables much better co-operation, help to share, exchange or even sell data.' However, by 2009 the Network had faded due to a lack of funding. Several member states wrote to the Commission to express their concerns. The new plant protection Regulation passed in 2009 obliges the Commission to report on the scope for a minor uses fund by the end of 2011.

Off-label approvals

Off-label approval provides a mechanism to make available approved plant protection products to deal with an economically damaging level of pests in a minor crop for which there is not a current on-label approval. One route has been the Long Term Arrangements for Extension of Use (LTAEU), introduced as a temporary measure in the late 1980s, which in effect allowed the use of pesticides on a minor crop providing that an approval exists for use of the same pesticide on a major crop. These arrangements did not fit in with the EU regime where specific approvals are required for all uses. PSD concluded that the LTAEU could not be maintained and those relating to edible crops were no longer valid from January 2007.

Growers or their representative organizations can apply for the approval of a pesticide on a specific crop under a Specific Off-Label Approval (SOLA). Such applications have to be supported by a reasoned case that covers the nature of the problem which must include reference to a named pest. Details must also be provided of the scale of the problem including its severity and scope and the potential for economic damage. One retailer argued in interview that more use should be made of the off-label approval route, stating 'You could use the retailer muscle to move that forward. That's where PSD and DEFRA should be looking.'

Maximum residue limits

MRLs are in effect a parallel system of regulation. The purpose of them is to ensure that residue levels do not pose unacceptable risks to consumers. It is important to emphasize that the MRL is not a health-based exposure limit, and exposure to residues in excess of an MRL does not necessarily imply a risk to health. Use of a pesticide would not be permitted in the first place if it led to exposure to pesticide residues above safety limits.

MRLs are established on the basis of the highest residue expected when a pesticide is applied in accordance with the approved conditions of use. MRLs applicable to UK trade are decided primarily at an EU level. However, although there are already more than 25,000 limits in force across the whole EU, member states also had the right to fix individual caps where the EU statute book is silent, and created over 100,000 according to European Commission estimates.

EC Regulation 396/2005 came into effect on 1 September 2008 to harmonize member state caps on pesticide levels. At the centre of the new rules is a phased-in harmonization of all MRLs specific to individual member states. All MRLs will either be substantive or temporary Community levels. The Commission will handle risk management, working on the basis of risk assessments from the European Food Safety Authority (EFSA). Member states will keep responsibility for enforcement, although the Commission will attempt to coordinate their efforts. A PSD official commented in 2008: 'MRLs will be harmonised, but it will be slow. It will get worse before it gets better.'

In the UK the Pesticides Residue Committee is an independent committee that supervises a programme of pesticide residues surveillance in food and drink at the point of sale with the work carried out on its behalf by the PSD. Generally very small numbers of samples are found to contain residues in excess of the MRLs. In the third quarter of 2007, 537 (71.7%) of 749 samples of 13 different foods had no detectable residues and 207 contained residues below the MRL level. The results showed that five samples (0.7%) contained residues in excess of the legal levels.

The practical difficulties that can arise from MRLs for growers can be illustrated by the case of dieldrin and courgettes. The use of dieldrin ceased in the mid-1970s, but it is still persistent in the soil. It can be extracted from the soil and concentrated in courgettes, pumpkins and water squashes, with uptake higher when the soil is particularly dry. Levels of uptake can vary significantly within one field. The residue issue is not one related to use, as there is no acute toxicology risk, but to environmental contamination and movement through watercourses. In 2006 the MRL level of 0.05 ppm was reduced to 0.02 ppm. In practice a number of plants are on or just above 0.02 ppm with a maximum of 0.03 ppm. From the grower's perspective, this created a situation in which the only way in which courgettes could be grown without breaching an MRL was hydroponically.

PSD reassured the grower that they would not take enforcement action, but the grower was concerned about the reputational damage resulting from adverse publicity. EFSA was to carry out a review of the toxicology of increasing the MRL. However, a PSD official commented: 'EFSA approach it in a very purist way. This was the separation of risk assessment and management. It goes back to the whole BSE story.' A more positive view of EFSA from within PSD was that 'EFSA is transparent, [it] forced [the] pace on transparency, getting Draft Assessment Reports into [the] public domain'.

In our interviews with retailers, some expressed the view that consumer misunderstanding of MRLs could lead to commercial damage. Retailers hire technical experts to check a sample of their produce to see if MRLs have been exceeded and take follow-up action with suppliers if this does happen. A number of retailers commented that problems were more likely to arise with imported produce, especially exotic fruit. One retailer commented: 'Growers are using far less chemicals than they have ever done, food has never been safer if only the public would realise that.' Another commented: 'Europe, dominated by green parties, focus[es] on MRLs. Have had a number of MRLs raised in recent years without any justification about whether it is right or wrong. We need to think about what we mean by residues.'

Enforcement and the 'black market'

Chemicals are among the most strictly regulated of substances because of their hazardous and toxic properties, and 'Pesticides are among the most strictly regulated of all chemicals' (House of Commons, 2005: 6). Pesticides 'emerge as a "special" risk regulation domain, characterized in terms of toxicological attributes, comparative strict regulatory controls and social benefits' (Blok et al., 2006: 321). '[The] data requirements that were fixed for pesticides greatly exceed those required for any other class of substance including pharmaceuticals, food additives and commodity chemicals' (European Commission, 2001: 3). When one has a strict system of regulation, a black market invariably appears. In the following discussion, a distinction is made between the 'black' market of illegally marketed products and the 'grey' market of products that fall outside the regulations but are competitors to biopesticides.

It has been estimated by the European Crop Protection Association (ECPA) that 5–7% of Europe's supply of crop protection products is fake. Criminal organizations are involved in this trade and '[t]he use of fake or illegal substances to treat crops is growing at an alarming rate' (Bounds, 2008). About 90% of the fakes come from China, with India being another source. At the British Crop Protection Council congress in 2005, lawyers acting for agrochemical companies served 37 injunctions on 19 separate Chinese companies. At the 2006 congress, one Indian and 23 Chinese companies were given legal warnings and three were ordered to cease commercial activity, after illegally promoting products. Ukraine has become an important crossroads in this trade. Differences in packaging may be very small and the first the grower may know that the product is not genuine is when the spray nozzle jams or the crop fails to respond in the expected way. Some of these sales may be made through the Internet, in other cases farmers have been targeted by text messages and in some cases the product comes from otherwise reliable distributors who have unknowingly purchased the material.

In late 2006 German authorities found isofenphos methyl, an illegal pesticide, in peppers from southern Spain. Raids led to the arrest of the president of the local cooperative and the seizure of 4000 kg of illegal pesticides among the greenhouses of Almeira, the centre of Spain's fruit and vegetable industry. However, the difficulties of dealing with such problems were emphasized by a retailer who discussed the Spanish case in interview: 'Technical managers fly out and try to get the bottom of it, more often you can't. Someone's touting an illegal, dealing with a central pack house with hundreds of growers in Almeira.' One consequence of these events was that pepper orders fell by a fifth as purchasers sourced elsewhere and prices halved. These economic effects had an impact on Spanish regional governments who encouraged more extensive use of biopesticides. Spain wants to transfer 38,000 ha of its vegetable production area to IPM with a focus on biocontrol by 2010 and is prepared to offer financial support of €1000/ha (Ehlers, 2007).

In 2008 PSD warned growers to watch out for a counterfeit herbicide labelled as 'Emrald Wotsit'. Nearly 3000 l of the counterfeit material had been identified, which had been manufactured on a different formulation from the approved Emrald Wotsit formulation. The formulation has been manufactured in Yorkshire and appeared to be originally intended for export, on which there are no controls. But the formulation had then been labelled as Emrald Wotsit for use in the UK, making it illegal to sell or use it, as well as constituting a breach of cross-compliance regulations.

Within the UK, PSD considers that the evidence points towards a small number of well-organized and persistent offenders, perhaps five or six. What makes things more complex is that not all the sales made are counterfeit products. 'The products targeted vary but are more likely to be older chemicals, which farmers are familiar with, but have had recent restrictions. An important feature is that they are usually for common pest and disease problems with high demand and high potential profits' (Abram, 2006). Prosecutions are very difficult and expensive because of the technical complexity of the cases. However, effective enforcement does not always require prosecution. PSD can issue enforcement notices which require remedial action to be taken. These can include monitored disposal of illegal products at the owner's expense.

The 'grey' market of biostimulants, foliar feeds and plant strengtheners is of particular concern to biopesticides manufacturers. Marketing products in this way is entirely legal so long as these products make no claim on their labels to control pests. Some of them are actually registered elsewhere as pesticides. It was estimated that as many 30 products were being sold as biostimulants in the UK in 2007, creating a way of going round the system of regulation. A manufacturer expressed the dilemma in the following terms:

> Because of the grey area around biostimulants and physical products, the grower does not feel confident of using biopesticides because of products sold with a wink and a nod. [If] the grower spends a couple of years using the product, then find[s] it doesn't work, [it] undermine[s] confidence in biopesticides. 'Muck and mystery' products. There is a lot of passing off going on; this product is as good as that one. The whole biopesticide product range is brought into disrepute.

In a 2009 paper the International Biocontrol Manufacturers' Association (IBMA) noted that some of its members were 'becoming increasingly aware that there are some products in the market place that are being described as "biopesticides" but which are not registered. What is not clear is whether these products comply with the legal requirements' (IBMA, 2009).

The IBMA saw the issue as falling into five categories.

1. Products containing microorganism active substances listed on Annex 1 (the main list of active ingredients for inclusion in plant protection products), being sold in the UK without national approval. While the regulations are isolate specific, the situation for additional examples of products of the same species is unclear.
2. Products containing active substances not included in the current EU review, with unproven efficacy, making claims of direct pesticidal activity.
3. Some products, particularly microorganisms, have both plant protection qualities and plant growth stimulant properties.

4. Products containing active substances, with known pesticidal activity, being sold without national approval as plant tonics, foliar feeds or other nutritional supplements.
5. Products incorporated into growing media, composts or as amendments that may include Annex 1 active substances or be considered as biopesticides.

In the Netherlands, a claim about plant strengthening falls under regulations involving indirect effect. All substances influencing plants are considered plant protection products. However, in Spain some 700 products are registered as 'phyto-fortificants'. Biopesticide producers consider that it is too easy to get under the radar with some products, but some grey market producers argue that they are unfairly targeted and are an important source of innovation.

Efficacy testing

The key efficacy issue is whether the product provides a measurable benefit when used according to its label recommendations. Efficacy testing raises a series of issues in relation to biopesticides. In the USA applicants for registration of biopesticides are not required to submit efficacy data, although they must have it available; but the USA is rather unusual in this respect in terms of national systems of regulation. Efficacy testing presents a number of challenges for biological products. With chemical products if a product works in the laboratory, it usually works in the field. Chemicals can be tested using quite small treatment plots but biologicals need larger plots to get statistically significant results because individual replicates are more variable: there is greater complexity and variation because they are living organisms. For the small and medium enterprises (SMEs) that produce biopesticides, the cost of efficacy trials can be a significant consideration. For a major crop or pest, ten trials are normally required, with three trials for a minor crop or pest. These would normally be undertaken over 2 years across a range of locations. A survey of 21 companies conducted for the REBECA project found that, across the EU, the average overall cost to achieve EU Annex 1 inclusion was €1.89 million, of which 21% was accounted for by efficacy tests (Hokkanen and Menzler-Hokkanen, 2007). One UK respondent claimed that efficacy testing could account for 50% of the registration costs for biopesticides as against 10% for chemicals. This figure is at the high end of the range, but another respondent suggested that costs in the UK of generating data were particularly high.

Why not let the market decide whether a product is efficacious? As a grower commented in interview, 'If a product was safe but we didn't know it would work we'd have a look at it'. PSD starts from the position that the Food and Environment Protection Act requires it to ensure that methods of controlling pests are efficient. It is therefore important to authenticate the claims made on the label about the product. This is a means of protecting users from deceptive claims about 'snake oil' products. A product must meet the EPPO Principles of Acceptable Efficacy so that it is significantly superior to the untreated control. In terms of Directive 93/71/EEC, efficacy data need to cover the justification of pest status of targets, the minimum effective dose and the use of officially recognized testing organizations. Firms developing biopesticides are not generally themselves 'officially recognized' to carry out their own efficacy trials. It is therefore important that they listen to advice from the officially recognized contractor and not insist on their own trial designs, which may be scientifically interesting but not necessarily what is required for registration purposes. A further category was recently introduced to the UK Official Recognition Scheme for 'Biologicals and Semiochemicals Trials/Tests'. This will allow organizations conducting work on microorganisms or semiochemicals to apply for official recognition in a more limited and specialized category of work than was previously possible.

Apart from the need to meet statutory requirements, PSD would argue that regulation can be seen as a cost–benefit analysis with the consideration of product efficiency being the benefit analysis. In its view efficacy is not an additional regulatory burden: the questions that PSD are asking are questions that applicants would be asking anyway to ensure that they have a marketable product. PSD would argue that it is flexible and pragmatic about sources of evidence, primary data, mode of action, resistance, dose response and glasshouse screening. There are no preconceived levels of effectiveness required for registration of a biological product. PSD can accept something with lower efficacy if it provides other benefits such as toxicological and environmental. If a worthwhile benefit can be shown, then a claim can be approved with an appropriately worded label to reflect the underlying data.

There is an issue about whether it is necessary to destroy crops used for trials. High-value crops can be worth up to a million pounds. The PSD view is that it cannot compromise safety by allowing products to go into the supply chain, but there may be further scope for flexibility on this issue.

Biopesticides Scheme

Concern about the lack of availability of biopesticides in the UK led to the introduction of the Pilot Project in June 2003 to facilitate their registration. Its aim was to increase the availability of biological pesticides in the UK by improving knowledge and raising awareness of PSD requirements and how to meet them. Prior to the introduction of the scheme, three active substances and four products were approved between 1985 and 1997. In April 2006 the Pilot Project was replaced by a Biopesticides Scheme. This covered four categories of products: (i) semiochemicals; (ii) microorganisms (bacteria, fungi, protozoa, viruses); (iii) natural plant extracts; and (iv) 'other' novel products on a case-by-case basis (no such cases had arisen by 2008). Since the introduction of the Pilot Project, seven products have been guided to approval. In April 2007 five products were at various stages of evaluation and several other companies were discussing possible applications with

PSD. Two products were approved in 2009 and several were at various stages of the registration process.

In order to operate the scheme PSD developed an internal biopesticides group made up of a team of specialists, including an assigned Biopesticides Champion, who understand the issues, have received specialist training and are continuing to learn. PSD thought it was desirable to involve as many people in an Approvals Group for this work as possible, rather than having a dedicated unit that would probably have insufficient work. These staff members are able to participate in pre-submission meetings with applicants. Particularly if they are held early in the process, they can help applicants to plan the acquisition of the data they need for registration and also avoid the compilation of any material which would be superfluous. In particular, it is helpful to hold a pre-submission meeting before efficacy trials take place.

A number of such meetings were observed on a non-participant basis as part of the research. A typical meeting might last for 3 or 4 hours with different experts from PSD joining the meeting as required. The meetings enabled the identification of gaps in the application dossier and mutually helpful discussions of how these could be filled, for example, through published data.

Reduced fees are charged for biological control agents: £22,500 for biopesticides; £13,000 for pheromones; and £7500 for taking either through EFSA procedures. These fees were not increased in March 2007 when other fees went up. Before the introduction of the Pilot Project, there was a standard fee of £40,000 for everything termed a biological. In comparison, the cost of core dossier evaluation, provisional approval and EFSA review for a synthetic would be between £120,000 and £180,000 from March 2007. CRD intends to continue to operate the Biopesticides Scheme with reduced fees.

The scheme has had to face a number of challenges. It has involved PSD reaching out to non-traditional 'customers' who may be suspicious of the regulatory authority because they have no experience of working with it. As a consultant commented, 'Pre-submission is a key element because registration is still an unknown, a lot of fear, people want me to hold their hands, introduce them to PSD'. For their part, applicants need to have realistic expectations. Putting an application in too quickly in the hope that the product will be available for the next harvest may reduce its quality. From a PSD perspective, the hope is that submissions will be of higher quality, needing less work and taking less time to process, thereby reducing the costs involved. An underpinning element of the PSD's work on biopesticides could be interpreted as giving it first mover advantage as the preferred registration authority for such products within the EU. It was claimed that 'Our biopesticides scheme is now a pathfinder in Europe' (Davis, 2008).

The European System of Regulation

This section explains the EU system of regulation as it operated in 2009, including a consideration of the way in which the review of actives is reducing their availability. Major criticisms of the EU system in terms of timeliness are considered.

It considers the problem of mutual recognition and possible solutions to it. The major revisions to the EU legislation which took place in terms of the review of Directive 91/414/EEC, its repeal with Directive 79/117/EEC and its replacement in 2009 by Regulation 1107/2009, along with the thematic uses strategy for pesticides and the framework directive on sustainable use are considered with particular reference to biopesticides. Reference is also made to the possible implications of the Water Framework Directive (2000/60/EEC) for which the implementing legislation came into force in 2004. There is then a consideration of two member states with very different systems of regulation, Denmark and the Netherlands.

The 91/414 European system

'Directive 91/414 was a new departure for the Community in several respects. It was one of the first major items of legislation to anticipate not only the principle of subsidiarity, but also the precautionary principle' (European Commission, 2001: 3). Its basic procedures remain intact in 2006/0136 (COD), but with some important modifications. The registration process is divided into two parts: Annex 1 inclusion of active substances and the authorization of plant protection products by member states. An extensive dossier is submitted by a company. It contains all the relevant information on the active substance and at least one representative product. It is submitted to a Rapporteur Member State selected by the company. The pesticide authority in that member state carries out a risk assessment and distributes the Draft Assessment Report (DAR) with a recommendation to the applicant and the other member states. In practice, 'a significant degree of variation became apparent between review practices among individual Rapporteur Member States. This problem had to be corrected by the development and adoption of a series of guidance documents' (European Commission, 2001: 5).

Since 2002 the EFSA has been responsible for risk assessment which it carries out through a scientific peer review by EFSA scientists and experts from the member states. This leads to the production of a guidance document which goes to the Working Group (legislation) of the European Commission's Standing Committee on the Food Chain and Animal Health. Member state civil servants decide whether to approve the active substance and, if successful, it would be added to Annex 1 of Directive 91/414. Product authorizations are considered at a national level using harmonized criteria for data requirements laid down in EU legislation (94/89/EC, Mammalian toxicology, including human exposure; 96/68/EC, Residues, including consumer exposure; 94/37/EC, Physical/chemical properties; 95/36/EC, Fate and behaviour; 96/12/EC, Ecotoxicology; 93/71/EEC, Efficacy). However, as one respondent commented, 'You have data requirements, member states can interpret them how you wish'. It should be noted that while microbials have a specific and uniform set of data requirements at EU level, there has been no specific set of data requirements for plant extracts, pheromones and semiochemicals. Chemical data requirements have been used, but cases could be made for not presenting part of the data.

Loss of active substances through the EU review programme

Many older substances are being removed from the market under the 91/414 review programme, either because they fail to meet the standards of the review or because companies decide for commercial reasons not to support them, which could include a consideration of the costs of re-registration. The MRL programme is likely to reduce the range of available uses. 'The EU review programme is having a significant impact on the availability of crop protection products. In some cases, there are no effective chemical solutions, as exemplified by the loss of all products for the control of root fly on swedes and turnips in the UK' (Richardson, 2005: 237). This does, of course, provide a new opportunity for biopesticides, but biological substitutes are not available for all the withdrawn products and, if they are, may be less efficacious.

The pesticide review was a protracted process and a new developed database finally came on stream in March 2009. About 1000 active substances on the market before 1993 were subjected to a detailed risk evaluation in terms of their effects on human health and the environment. Two hundred and fifty achieved the harmonized EU safety assessment and 70 substances failed it and were removed from the market. The others were eliminated as dossiers either were not submitted or were withdrawn by industry.

Shortcomings in the EU system of registration

The length of time it takes to complete registrations is a major consideration for SMEs as they wish to generate a cash flow as soon as possible and start to recoup the costs of research and development. 'The industry complains that the current registration period for [microbial biological control agents] in the EU is costly and time-consuming. Long registration periods are a severe problem, because they delay the onset of the returns for the investments made during research and development' (REBECA, 2008: 19). Seven examples of biological control agents considered by the REBECA project showed that the time period from the month of dossier submission to the month of inclusion on Annex 1 or granting of national registration was a mean of 87.7 months with a range of 57–121 months (REBECA, 2008: 19). The mean time for registration in the EU was 75 months compared with 28 months in the USA (Hokkanen and Menzler-Hokkanen, 2008). There was also considerable variation in reported registration times between member states. Britain and Germany were regarded by our respondents as having relatively efficient registration agencies. A number of complaints were made about Spain, one respondent commenting, 'waiting two years from submission, such a backlog of products'.

In an internal market, one would hope that once a product was approved in one member state, it could then easily be registered in other member states. As one manufacturer commented, 'Hopefully, once done a registration in one country, just a rollout to other European countries. Everyone is looking for mutual recognition routes. Have a lot of trouble with mutual recognition.' The Commission accepts that 'Mutual recognition does not function well and national authorisations of products leads to duplication of work in Member States and to differences in the availability of plant protection products across the European Union' (European Commission, 2005).

Some of these problems have arisen because of the existence of 27 regulatory authorities across the EU with varying and often inadequate levels of resources available to them and following their own particular procedures. Some member states are uneasy about the principle of mutual recognition. A Danish Environmental Protection Agency (EPA) official commented in interview: 'Mutual recognition would seek to undermine standards. We really feel that.' The slowness of getting Annex 1 listings has also been a factor and the drive to speed things up led to substances being approved for single rather than multiple uses. A more underlying tension is, as one respondent put it, 'Regulators united in commitment to carrying on regulation on a member state basis, policy makers likely to favour a more harmonised approach'.

To some extent, regulators have been developing their own informal links, both through EU and OECD committees and on a bilateral basis. One participant in the EU expert meetings noted that Belgium, Denmark, Germany, the Netherlands, Sweden and the UK were regular in attendance with France and Hungary coming occasionally. These links can be used as a basis for information exchange and have led to work sharing of the Rapporteur role. A PSD official commented, 'Try to get an air of pragmatism, work sharing, where we can support each other, where we can support each other to cut down on data requirements. We used to know frighteningly little about how others did efficacy.' PSD sees itself as 'trying to push pragmatism into Europe. [We are] trying to push pragmatism into other member states.'

One available mechanism for harmonization is to make use of the EPPO. Founded in 1951, EPPO is an intergovernmental organization with 50 members and a secretariat of 12 based in Paris, funded by contributions from the member governments. In the plant protection area, EPPO has developed a number of standards for use by national registration authorities. PSD has taken certain UK procedures and has turned them into EPPO ones, particularly in the efficacy area, for example trials guidelines and minimum effective dose. One respondent suggested that EPPO had more effective dissemination mechanisms than the EU.

Although the EU is supposedly an internal market, it is geographically diverse and pesticides that work in one set of climate conditions will not necessarily work in another or at least not as well. A proposed solution has been to devise climatic zones in which mutual recognition could occur. This idea originated with a paper by the UK Crop Protection Association that outlined a northern European region with a climate comparable to the UK. This paper was accepted by the ACP and 'provided an opportunity for companies to use northern European data in regulatory submissions without the need to provide for each trial a specific justification that climate conditions were relevant to the UK' (Richardson, 2005: 235). More generally, 'The climate within a zone can be considered comparable for the purposes of effectiveness data, and should aid the acceptance of efficacy data between countries within a zone' (Richardson, 2005: 235). However, as will be evident from the following discussion of EU reform proposals, zoning met with considerable political resistance and many problems connected with its implementation have to be resolved.

The reform of the EU regulatory system

This revolves around three main proposals: the amendment of Directive 91/414; the EU Thematic Strategy; and the Sustainable Use Directive, but also discusses the Water Framework Directive. In this section the main elements of each proposal are considered and then their progress through the co-decision process of the EU. There is also a draft regulation on the collection of data on pesticides sale and use, but that is not considered further here.

The amendment of Directive 91/414

A steer for the revision of the directive was provided by the 2001 report to the Parliament and the Council on how the directive was working. It was admitted that 'Decision-making was slow between 1993, when the Directive entered into force, and 1999 … An already complex piece of legislation became progressively more complex in application as expectations mounted on all sides, ensuring that standards and criteria were not only maintained at a high level but were effectively raised' (European Commission, 2001: 8). It took another 5 years for the Commission to produce a set of proposals for amendment. Part of the reason for the delay was no doubt the extensive consultations held with a range of stakeholders, including the IBMA. The other principal stakeholder groups consulted were: agriculture and food organizations, six; agrochemical and chemical organizations, four; environmental organizations, four; animal welfare organizations, two; retailers, one; consumers, one. This gives the indication of the range of conflicting interests involved and helps to explain why it was so difficult to get the proposals through the Parliament. Despite the time taken in the preparation of the proposal, one regulator commented, 'It is not the best thought out proposal. Born of frustration that mutual recognition has not taken off.'

A central consideration for the Commission was that 'The large difference in authorisation of existing active substances shows that without further harmonisation the protection levels in Member States may vary a lot' (European Commission, 2006a: 9). The whole approach of the document was informed by the application of the precautionary principle to 'ensure that industry demonstrates that substances or products produced or placed on the market do not adversely affect human health or the environment' (European Commission, 2006a: 14). The document also endorsed the principle of IPM, but recognized that it would require a transition period of unspecified length before it could be applied, although the accompanying Thematic Strategy targeted 2014.

The main proposals included provisions to encourage a move from national to zonal-level authorizations with obligatory mutual recognition between member states of the same zone. This solution to the mutual recognition problem had proved particularly controversial during the stakeholder consultations and continued to be so subsequently. Some stakeholders opposed any further harmonization, while others favoured full centralization. 'The proposed system is a compromise between the current situation and a fully centralised authorisation' (European Commission, 2006a: 7). Like many compromises, it pleased relatively few with only Austria and the UK being clear supporters among the member states with France and Germany opposed. One consideration was that states with national testing systems, such as France, did not want the job losses associated with greater centralization.

A key plank of the proposals was risk-based comparative assessment and substitution to encourage the replacement of more hazardous substances and/or products with alternatives. This was good news for advocates of biopesticides, and has been called for by the IBMA in its 2005 'white paper' on the regulation of biologicals, but the influential ECPA was opposed in principle to regulatory substitution at any level. Farmer organizations were also opposed because they were concerned about the reduced availability of products. Hazard triggers would be used to exclude from Annex 1 carcinogens, mutagens or reproductive toxins.

The Commission proposed bringing to an end the system whereby member states could grant provisional national authorizations for new active substances pending a decision on Annex 1 listing. This was thought to provide a welcome element of flexibility in the system, but the Commission's view was that it was incompatible with a harmonized system for establishing MRLs and would be offset by strict deadlines for assessing an active substance. Whether deadlines such as completing the DAR within 12 months of notification of the completeness check are achievable is another matter. Resource constraints and the availability of specialized personnel have been a continuing constraint for the European pesticides regulation regime.

The Thematic Strategy for Pesticides and the Sustainable Uses Directive
This arose from the 6th Environmental Action Programme, which called for a reduction in the impact of pesticides on human health and the environment and the need to achieve a more sustainable use of pesticides. However, the strategy document acknowledged that 'The potential risks associated with their use are accepted to a certain extent by society given the related economic benefits since *inter alia* plant protection products contribute to ensuring supplies of affordable and healthy agricultural products of high quality' (European Commission, 2006b: 3). The Commission noted with concern that 'Despite the increasing costs involved in [the registration] process and the decreasing number of active substances on the market, actual consumption and use of pesticides in the EU have not decreased in the last ten years. At the

same time, the percentage of food and feed samples where residues of pesticides exceed maximum regulatory limits is not declining, but remains around 5%' (European Commission, 2006c: 3). As with the proposal to amend Directive 91/414, the Commission expressed its concern about policy variability between member states, which in their view required harmonizing action at EU level: 'Over the last fifteen years, significant but uneven changes in pesticide use have been observed amongst Member States. Whilst pesticide use declines in some Member States, a sharp increase has been observed in others' (European Commission, 2006c: 3).

The Commission declared its support for forms of agriculture that restricted or better targeted 'the use of plant protection products, such as organic farming, integrated pest management or the use of less susceptible varieties' (European Commission, 2006b: 6). No mention is made of biopesticides, reflecting their low profile at EU level. This is evident in internal SANCO documents which examine extensively the reaction of the ECPA to proposals, but make no mention of the IBMA. During the debates in the Parliament, lobbying efforts by the IBMA raised the profile of biological solutions, but the toehold they obtained in the legislation was subsequently removed.

The preparation of the Thematic Strategy identified the need for a Sustainable Use Directive as several of the envisaged measures could not be integrated into existing legislation or policies. This directive was passed in 2009 and will be implemented by 2011. It was centred around the creation of National Action Plans to identify areas of risk, the reduction of risk and use, the minimization of impacts on human health and environment, responsible use and IPM. It would set up systems of training for distributors and users, require member states to set up systems for inspection of application equipment, and prohibit aerial spraying.

The battle in the Parliament
In its plenary vote in November 2007, the Parliament substantially toughened up the proposals put forward by the Commission. As they stood, the Commission proposals on active substance approval criteria would remove the main control for major diseases of wheat in the UK with potential yield losses of 20–30% and would have significant implications for crops such as carrots, parsnips and onions because the majority of approved herbicides might no longer be available. The Parliament proposals would mean 'effectively no herbicide options for control of weeds in horticultural crops; chemical control of black-grass in cereals would become virtually impossible with severe economic impacts … potatoes – seed potato growing unlikely; ware potato yields severely reduced … many horticultural crops would be uneconomic to grow' (PSD, 2008b: 11–12).

The discussion of the Parliamentary debate focuses on two proposals that are particularly relevant to biopesticides: the zoning scheme and an amendment proposing distinct regulatory arrangements for biopesticides. The scheme to divide the EU into northern, southern and central zones for mutual recognition purposes had attracted opposition as soon as it was proposed. There were concerns that the proposed zones did not represent homogeneous regions in terms of climate, pests or agricultural production and practices, although truly homogeneous zones would be difficult to devise. Hungary in particular pressed for a fourth zone which would presumably cover Eastern Europe. There was concern that products might be withdrawn from the market unnecessarily because of a conservative assessment by one member state. Denmark, which takes a relatively rigorous stance on pesticide regulation, expressed concern that a plant protection product 'approved in one Member State could have a highly adverse effect in other Member States of the same zone' (Anon., 2007). From a different perspective, an industry spokesman commented that 'Introducing a zonal system often takes away national flexibility, instead of a lowest common denominator system which is often talked about in Europe, going to end up with potential for highest common denominator in granting an authorisation. Any member state that has an issue with something imposes that condition.'

It was always possible to criticize the arbitrary nature of the zones, particularly given that it was not politically feasible to divide countries as might have been climatically sensible for northern and southern France. German Rapporteur Christa Klass complained that the scheme would lead 'to the introduction of a zoning line which would, for example, place the right and left banks of the Mosel in different "legal areas"'. She therefore suggested that instead of arbitrary zones, that the principle of mutual recognition of national licensing should be retained, but that the Member States should, in the spirit of the subsidiarity principle, be allowed to make national or regional specifications' (European Parliament, 2007: 132). In other words, there would be a completely discretionary system in which member states could opt to make a regional ruling, although it is unclear how other member states might then respond. The European Parliament subsequently rejected the zonal authorization system and the linked compulsory recognition of authorization within a zone. The Commission rebuffed this rejection: 'The amendments have not been accepted as they would have considerably undermined the Commission proposal and would have removed one of its key elements' (European Commission, 2008: 7). However, in an effort to reach agreement the Commission put forward a compromise 'in the form of "exceptional derogations" in a bid to render the authorisation process more relevant to realities on the ground such as climate conditions, soil types, crops cultivated etc.' (Anon., 2008).

The plant protection regulation created northern, central and southern zones and a single zone for greenhouses, seed and post-treatments, empty rooms and containers. Initial authorization will be undertaken by a member state in the same zone, but application can be made in any member state for greenhouses. In general the application is examined by the member state proposed by the applicant. Other member states in the zone would normally recognize the authorization decision, but if they have human or animal health or environmental concerns, and there are no risk mitigation measures, they can refuse to authorize. How mutual recognition will work in practice remains to be seen. The whole procedure for the inclusion of an active substance should take 30 months, plus 15 additional months if additional time periods and clock stops are used.

Rapporteur Christa Klass proposed a specific system of regulation for biopesticides. In order to meet substitution objectives, 'the licensing of natural biological control products such as micro-organisms and viral and pheromone products should be facilitated. When the relevant research projects have been completed, these substances should not fall within the scope of this regulation, but be specifically regulated' (European Parliament, 2007: 132). In Amendment 35, the European Parliament inserted language giving priority to non-chemical alternatives for plant protection. The Commission deleted this amendment, justifying its rejection in the following terms:

> Amendment 35 aims to introduce a future limitation to the scope excluding micro-organisms, viruses, pheromones and biological products once a specific regulation related to these products will be adopted. The Commission retains that there is no need for a specific regulation as specific data requirements and criteria for authorisation are already in place and for some of these substances the provisions concerning low risk substances could apply.
>
> (European Commission, 2008: 5)

The new legislation does give a specific status to non-chemical and natural alternatives and requires them to be given priority wherever possible. The definition of non-chemical methods mentions in Article 3(8) 'physical, mechanical or biological pest control'. Biological controls should generally qualify as low-risk active substances in Annex II, Section 5. They should certainly satisfy the exclusion criteria in that they are not carcinogenic, mutagenic, toxic to reproduction, toxic or very toxic, nor are they sensitizing chemicals, explosive, corrosive or deemed to be endocrine disruptors. Problems may be caused by the requirement that they are not persistent in the sense that their half-life in the soil is less than 60 days. This provision will almost certainly require a guidance document as provided for in Article 77 which permits the Commission 'to adopt or amend technical and other guidance documents ... concerning micro-organisms, pheromones and biological products, for the implementation of this regulation'. EFSA would normally undertake this work, but its schedule is full for 2010 and probably for 2011 (Heilig, 2009).

Being classified as a low-risk active substance gives initial approval for 15 years rather than the standard 10 years. A fast track decision on authorization should be taken within 120 days providing no data are missing. A reduced dossier can be submitted for plant protection products containing them, but this has to include a demonstration of sufficient efficacy. The products may be advertised as 'Authorised as low-risk PPP', but paradoxically no such claim can be made on the label. It should also be noted that the final categorization of low-risk substances will depend on EFSA peer review.

The new European legislation does not give biological controls all they might have hoped for, but it does give them legislative recognition and opens up the potential for faster authorization processes and effective mutual recognition.

The Water Framework Directive
The implementation of water quality legislation in the form of the Water Framework Directive could also have important implications for the availability of active substances which could go beyond those anticipated as the result of the revision of 91/414. If certain pesticides were found to be continually damaging water quality, they could be withdrawn. The active substances that are most likely to be affected are those used on a large area and/or at high rates. This is typical of herbicides used on combinable crops where ten herbicides are causing some concern, in particular those used for the control of grass weeds in oilseed rape. Many insecticides could also be at risk, making options for control of some pesticides very limited. Metaldehyde is particularly at risk as it is being found in water and this could lead to difficulties in slug control.

Denmark: regulatory stringency
'Scandinavian regulatory standards are widely believed to be stricter, less favourably disposed to pesticide usage, than is the case in countries like the UK and parts of Southern Europe' (Blok *et al.*, 2006: 313). 'In particular, Denmark is seen by most experts, including the NGO respondent, as possessing comparatively restrictive regulations, as opposed to countries such as Italy where "south of Rome normal rules do not apply"' (Blok *et al.*, 2006: 319). This was confirmed in interview by a Danish EPA official, who stated: 'There are some areas where we are more strict than other member states. For most effects we use a high safety factor.' Apart from a general 'green' orientation, a major specific driver for this stance is that 'Contrary to most other European countries, Danish water supply companies use groundwater reserves for drinking water, with no or minimal forms of prior purification' (Blok *et al.*, 2006: 314). Denmark may, however, be less of a regulatory leader on pesticides than it was. Following a change to a more conservative government in 2001, 'concerns are raised that the content and style of environmental policy making may be changing' (Blok *et al.*, 2006: 315). In particular, the Danish EPA 'used to be populated by environmentally committed staff unwilling to approve pesticides' (Blok *et al.*, 2006: 317), but this is no longer the case.

Denmark is particularly noted for its use of a pesticide tax, with Norway being the only other European country to use such a policy instrument, In 1986 the Danish parliament approved a pesticide action that required a 50% reduction in the use of pesticides by 1997. Use decreased less than had been hoped for and in 1995 a tax was imposed 'equal to 37 per cent of the retail price on insecticides and 15 per cent on fungicides, herbicides and crop-regulating chemicals' (Daugbjerg and Pedersen, 2004: 225). In 1998 the tax was doubled but the government had to back down in the face of pressure from agricultural interests and lower land taxes to compensate. In Denmark '[t]he combined sale/consumption of plant protection agents to/in agriculture has fallen from 6,972 tonnes of active substance in 1981–1985 to 2,889 tonnes in 2000 ... The experience gained from pesticides taxation shows that the tax – in conjunction with a range of other measures – does have an effect' (Larsen, 2004: 18). However, it should be noted that these figures take no account of the toxicity of active ingredients. In Britain the New Labour government considered the use of a pesticide tax after it came into office in 1997 but decided to rely on the Voluntary Initiative to minimize the impact of pesticides on the environment and on water quality in particular.

Does the Danish approach facilitate the wider use of biological controls? The Bichel Committee, which was set up by the Danish Government in 1997 to provide an overview of the use of pesticides, devoted less than a page to biological control and adopted a somewhat sceptical tone:

> The utilisation of useful organisms and microbiological methods would ... present a significant risk of proliferation of alien organisms, which could have a detrimental effect on the environment. Theoretically speaking, the proliferation of local species could also disturb natural ecological balances. The use of microbiological methods entails a risk of industrial injuries, in the form of allergies or industrial diseases.
> (Bichel Committee, 1998: 73)

Biopesticides are subject to the pesticides tax, but at a lower rate of 3%. However, as there are no fees for authorizations and applications, the low tax is intended to cover administrative costs instead of a fee. If Denmark is a Rapporteur Member State and has to do an EU evaluation, a charge of €100,000 is made for microbials compared with €210,000 for chemicals. Research is funded from the pesticides tax and, as far as is possible given the quality of the applications received, one microbial project is funded each year. For a time, the Danish EPA lacked an expert to deal with biological controls: '[The] person responsible stopped work at EPA and [products] were put on the shelf for many years. Then another person came and not much happened, then I got the job.' This does not suggest a very high priority for biological controls. The emphasis of Danish policy seems to have been on reducing pesticide use rather than promoting alternatives.

The Netherlands: regulatory innovation
Denmark does not have a large protected crops sector, but its importance in the Netherlands has created a favourable environment for the promotion of biopesticides. Dutch Government policy is concerned with reduction of use, but also covers the stimulation of the use of natural products. Dutch law has a provision for products with a low toxicity profile (regulation exempted pesticides, RUB). Genoeg is a series of schemes which have attracted financial and political support from the Ministry of Agriculture. Genoeg is the acronym for Gwasbeschermingsmiddelen van Natuurlikje Oorsprong Effectief Gebruiken, which translates as 'using plant protection products of natural origin more effectively' (or, more colloquially, the effective use of natural pesticides). It is a functional equivalent of the UK Biopesticides Scheme, although in the Netherlands companies receive a subsidy from public funds while in the UK fees are lowered. In the Netherlands, the process was a more 'bottom-up' one than in the UK with a coalition of actors creating a new process.

The first phase of the project from 2001 to 2002 was an exploratory one concerned in particular with the development of an inventory of effective natural pesticides. After a phase of issue management concerned with the dissemination of knowledge to growers, Project Genoeg Toegelaten ran from 2003 to 2005 and was concerned with the support of five natural pesticides for use in glasshouses. One product approved through the Genoeg scheme was not approved in the UK by PSD. From 2004 to 2008, Project Genoeg Breed focused on the support of ten natural pesticides for all uses. In 2008, seven agreements for support were in place involving the co-financing of 13 new products.

The project is managed by a consultancy called the Centre for Agriculture and Environment (CLM), which has extensive experience in sustainable agriculture. It was started to serve as an intermediary between farmers and environmentalists and has good links with all stakeholders, cemented by regular meetings. The consensus and coalition building has been a key element in the success of the project. In selecting products for support, an emphasis is placed on products that have a low toxicity profile and where data on efficacy and toxicity are available. Products are ones where successful authorization is expected and where the company will defend the product.

Fifty per cent of the costs of research and registration are co-financed to a maximum of €100,000. Help is provided in literature search for registration by a body that has expertise in technical questions about ecotoxicology, the National Institute for Public Health and the Environment (RIVM). The regulatory body, the Board for the Authorization of Pesticides (CTB), provided a help desk facility to assist applicants and established a biopesticides team made up of six staff members each with a defined expertise or speciality such as characterization or residues.

There have been some problems with companies dropping out during the process. For many applicants, the cost of registration remained a problem with €400,000 being identified as a typical cost for the whole process. As in the UK, a key lesson has been the importance of meetings with applicants early in the application process to identify areas of difficulty and possible solutions.

The USA: Promoting Biopesticides

The USA has an impressive record in promoting the registration of biopesticides. Since current data requirements were published in 1984, nearly 80 microbial pesticides have been registered: 39 bacteria; 29 fungi; seven baculoviruses; two yeasts; and one protozoon. The success of the US EPA's efforts to give priority to the evaluation of reduced-risk pesticides is illustrated by the fact that in 2004, of the 26 new active ingredients registered, over half (14) were biologicals and five were reduced-risk conventional pesticides (Lindsay, 2005: 341). The EPA has 'always had a traditional pesticide regulatory process, which is data rich and resource intensive' but is seeking to move from 'data heavy to data smart' (Lindsay, 2005: 343). In particular, '[w]hat is needed is a shift from a paradigm where extensive hazard testing is conducted, and information that is not relevant is eliminated, to one where a risk-based, hypothesis driven approach identifies the information most relevant to the assessment' (Lindsay, 2005: 345). The effort to promote biopesticides in particular depended on a few committed individuals: 'Fifteen years ago nobody paid us any attention'. What changed this indifference was that '[t]here were staff who were really interested, senior leadership at a political level within the Agency. Upper management tell us, spread the technology, get it out there.' Like the PSD, EPA provides pre-submission meetings which discuss how to satisfy data requirements.

Direct comparisons between the USA and the EU are inhibited by the fact that regulation in the USA is largely undertaken at the federal level, whereas the EU has a complicated division of responsibilities between the European institutions and the member states. Pesticide labelling is entirely a federal matter in the USA, but as confirmed by a 1991 US Supreme Court judgment, *Wisconsin Public Intervenor v. Ralph Mortier*, federal pesticide law does not pre-empt local regulations dealing with the use of pesticides. A state can refuse to register a product and hence the sale and use of any pesticide. The two most active states in the area of pesticides regulation, both of which have large fruit and vegetable sectors, have been California through its Department of Pesticide Regulation and Florida through the Florida Pesticide Law. California claims to have the nation's toughest pesticide laws and the largest and best-trained enforcement organization, with about 350 state inspectors compared with about 45 in Florida. The state has been particularly active in the encouragement of IPM. Unlike the federal regulator, California does require efficacy data, although there are exemptions if it has previously approved products containing the same active ingredient. At the federal level, applicants for registration are required to have efficacy data available, but they are not usually called in.

Nevertheless, the main impetus comes from the federal level. The EPA has a dedicated Biopesticides and Pollution Prevention Division, which is the smallest division in the Agency and in a different location, both factors that may have contributed to the proactive philosophy and committed culture observed when the Division was visited as part of the research. The Division head commented in interview: 'We believe strongly that there's a role to create a division just dedicated to [biopesticides]'. Another staff member commented:

> We do things differently from the rest of the office; we have a one-stop shop, don't farm out anything on economic or ecological efficacy or health effects. We have our own chemists, economists, toxicologists; cuts down on amount of bureaucracy, helps us to implement things. Have all personnel in one division, on [the] same floor, much easier if [the] person is sitting in [a] cubicle across [the] floor.

This point was reinforced by the observation that legal experts from the office of general counsel visited just once a week: 'That's where the biggest friction is, would be easier if full-time; does help familiarity'.

EPA staff members attach considerable value to the Interregional Research Project (IR-4) programme, which operates as a partnership between the United States Department of Agriculture (USDA) and the land grant universities. This was started in 1963 and remains focused on minor crops rather than biopesticides, but started a Biopesticides Program in 1982. The programme was expanded in 1994 'to provide competitive grant funding to support studies on early development stage biopesticides for minor crop uses' (Holm *et al.*, 2005a: 241). IR-4 helps biopesticide registrants with regulatory advice and the preparation of applications for approval. It also developed organizational capacity by supporting the growth of the Biopesticides Industry Alliance (BPIA) as a trade association of biopesticide companies. Biopesticides division staff work closely with IR-4 and it was noted in interview: 'As a government agency we're prevented from promoting a product or technology'.

Comparing the European and US systems, a consultant commented: 'The US system, in terms of data requirements is not less onerous but more accessible to applicants because of the biopesticides sub-unit of EPA, provides list of data requirements'. Regulators in Europe were a little more sceptical about the American arrangements. One biological product examined by PSD turned out to have a human pathogen in the formulation, although the product had been approved in the USA. A regulator commented: 'Reduced risk means all things to all men in EPA. Difficult to identify from legislation what is reduced risk.'

In terms of the number of products registered, the US system has been more successful than that in Europe. This may not be just due to differences in the regulatory arrangements, but also to the existence of a large internal market covering a variety of climatic zones that allows development costs to be recouped relatively rapidly. The fact that the European market is fragmented by regulatory barriers is a reflection of shortcomings in the realization of the European project.

International Harmonization: the Role of OECD and WHO

It has been argued that 'the development of harmonized guidelines and terminology by the OECD is critical to the advancement of work sharing on the international level' (Holm *et al.*, 2005b: 31). The Biopesticides Steering Group of OECD, which was established in 1999, provides an important forum for regulators and industry to meet and work together. Working together enables a speedier and more thorough evaluation of a biopesticide's risk, thus (it is hoped) speeding up the process of approving safer pesticides. The OECD work has focused on writing guidance for the submission and evaluation of test data. The agreed guidelines establish two formats, one for industry to use when submitting dossiers and one for regulators to use when writing their evaluation reports. The OECD guidance is intended to ensure that dossiers and evaluation reports are clear and complete, and that information is easy to find. This makes it easier for governments to use each other's pesticide risk evaluations, overcoming potential non-tariff barriers. The practical value of the guidance is shown by the fact that they were used in relation to pheromones in the pilot scheme. One manufacturer noted that the OECD guidelines were in some respects more arduous than what is required by some national regulatory agencies.

Not all countries are active participants, with Japan reportedly having never attended. There was a feeling in Europe that 'the US try to keep much of their own system rather than compromising'. This was echoed by an EPA respondent who commented: 'I think we can harmonise with Canada, but I am not sure about all the other countries of the world'. The North American Free Trade Agreement (NAFTA) has a Technical Working Group on biopesticides and agencies in Canada, Mexico and the USA undertake joint reviews using OECD guidelines.

The WHO has had a Pesticide Evaluation Scheme (WHOPES) since 1960 that promotes and coordinates the testing and evaluation of pesticides for public health. This covers the safety, efficacy and operational acceptability of pesticides and develops specifications for quality control and international trade. In 2001 the FAO and the WHO agreed to develop joint specifications for pesticides.

Design Principles for a Better Regulatory System for Biopesticides

As part of our research, we sought to develop a set of design principles for a better regulatory system for biopesticides. Our hope was that these principles could inform future policy debates and assist the coherence of regulatory proposals. We developed five underlying principles.

1. Biopesticides have a key but specific role to play in crop protection. Synthetic chemical pesticides are a precious resource for crop protection and should be treated as such, but problems of resistance and the withdrawal of some products mean that fewer chemical pesticides are available on the market. Alternative crop protection tools are needed and biopesticides have an important and increasing role to play in crop protection, although normally in combination with other tools including chemical pesticides as part of IPM.
2. Biopesticides should be regulated. Just because something is natural does not mean that it is safe. However, the regulatory system has tended to follow a chemical pesticide model that does not facilitate the efficient registration of biopesticides.
3. The regulatory system must support sustainability objectives. It should be informed by a proper understanding of the precautionary principle. While there should be a special emphasis on environmental sustainability, other aspects of sustainability including economic sustainability should be taken into account.
4. Pest management should be ecologically based. It should make use of natural enemies of pests and be undertaken within the context of Integrated Crop Management (ICM) and IPM. Biopesticides offer benefits to both conventional and organic farming, providing an aspect of unity where there is otherwise division.
5. It should be credible with all stakeholder groups including consumers. It is particularly important that the regulatory system has the confidence of consumers in relation to human health issues.

Operational principles

There needs to be an improved knowledge basis and chain. There still needs to be better understanding of the ecology of microbial control agents, the fate and behaviour of microbial releases, and of the effects of secondary metabolites/toxins produced by microorganisms, a minority of which could have consequences for human health.

Stakeholders need to be fully involved in the debate on regulation and its implementation and to have effective communication links with each other. As is discussed further in Chapter 6, the biopesticides policy network is weakly developed and sometimes politically unsophisticated. A key problem is the lack of an influential constituency of support for biopesticides, particularly in environmental terms with the debate about a more sustainable agriculture being framed around discourses about organic farming. The role of retailers in imposing additional requirements on growers poses a number of challenges that are discussed more fully in Chapter 7.

There needs to be a biopesticides 'champion' organization, preferably quasi-governmental in character, which can act as an advocate for biopesticides. In the USA this role is to some extent performed by the biopesticides division of the EPA, but the American model cannot simply be transferred to Britain. Such a role would be compatible with the broad task and objectives of Natural England in the UK, but it is not a role that it is ready to adopt. Nevertheless, without an organizational champion with a clear advocacy role, the case for biopesticides risks being sidelined.

Whether the organizational absorption of PSD within HSE will be beneficial remains to be seen, but it is important that the York location is maintained to avoid the loss of experienced staff and consequent disruption to the biopesticides programme. PSD has led the field in Europe in biopesticides regulation through the Biopesticides Scheme. The existing staff network dealing with biopesticides within CRD should continue to be supported and developed in terms of enhanced coordination and training, a clear group identity and strong organizational support. ACP needs more knowledge on biopesticides in general and the biology of microbial control agents in particular. This could be achieved by creating a network of experts with appropriate specialisms.

Although regulators in Europe support efficacy testing, views within industry tend to be somewhat divided. It is in any case necessary for marketing claims and it protects users against deceptive claims from 'snake oil' producers who can damage the reputation of an emergent industry. There may be scope for varying some of the requirements imposed in efficacy testing and hence reducing the cost. These issues are being considered by the OECD Biopesticides Steering Group. Consideration should be given to the findings of the EU-funded Efficacy Extrapolation Project carried out by PSD. A report was made which presents proposals for harmonizing efficacy and crop safety extrapolations with the aim of reducing the need for trial data while maintaining high standards of performance and selectivity. A key area for review is the number of tests required over what time period.

A number of important developments are taking place at the EU level. The role of the EFSA is still evolving, but concerns have been expressed that its application of the rules may be too inflexible and more data will be requested for a risk assessment than are absolutely necessary. A key requirement is to ensure that mutual recognition works effectively between member states. This would create a larger market for biopesticides and overcome some of the economies of scale problems that manufacturers face. The ecozone proposal has attracted some criticism, but other than the further development of cooperation between national registration agencies, it represents the most feasible way forward.

Considerable sums of public money have been spent on the development of biocontrol products which have not then been registered or marketed. SMEs have encountered considerable difficulties with the cost of the registration process, not just the registration fee, but the cost of development and testing. Data collated by David Richardson about government direct or indirect funding across 14 European states show that only three member states provide no funding: Estonia, Slovenia and the UK (Denmark offsets fees against the pesticides tax). Six countries, including France and Germany, provide funding for trials. Such assistance may be seen as too 'near market' in the UK, which tends to follow a relatively rigorous definition of market failure, whereas the USA is more inclined to support new technologies seen as serving the national interest. There is a market failure to be addressed in terms of the positive social and environmental externalities of biopesticides not being realized.

References

Abram, M. (2006) Too good to be true: the pesticide scam. *Farmers Weekly*, 7 April, 16–17.
ACP (2004) *Alternatives to Conventional Pest Control Techniques in the UK: a Scoping Study of Their Potential for Wider Use*. Advisory Committee on Pesticides, York, UK.
Advisory Committee on Pesticides (ACP) (2005) *Crop Spraying and the Health of Residents and Bystanders: a Commentary on the Report Published by the Royal Commission on Environmental Pollution in September 2005*. Advisory Committee on Pesticides, York, UK. Available at: http://www.pesticides.gov.uk/acp_home.asp (accessed 13 July 2010).
Anon. (2007) Any other business – plant protection products and sustainable use of pesticides. *Agra Focus*, July issue, 11.
Anon. (2008) May Council political agreement on pesticides now likely. *Agra Focus*, May issue, 19–20.
Bichel Committee (1998) *Report from the Main Committee*. Danish Environmental Protection Agency, Copenhagen.
Blok, A., Jensen, M. and Kaltoft, P. (2006) Regulating pesticide risks in Denmark: expert and lay perspectives. *Journal of Environmental Policy and Planning* 8, 309–330.
Bounds, A. (2008) Alarm at flood of bogus pesticides. *Financial Times*, 20 May.
Daugbjerg, C. and Pedersen, A.B. (2004) New policy ideas and old policy networks: implementing green taxation in Scandinavia. *Journal of Public Policy* 24, 219–249.

Davis, R. (2008) The regulatory response: the Biopesticides Scheme. Paper presented at Workshop on *Biopesticides: The Regulatory Challenge*, Warwick HRI, Warwick, UK, 31 October.

Ehlers, R.-U. (2007) Perspectives for safer plant protection. Paper presented at the *REBECA Conference on Balanced Regulation for Biological Plant Protection*, Brussels, 20–21 September.

European Commission (2001) *Report from the Commission to the European Parliament and the Council. Evaluation of the active substances of plant protection products (submitted in accordance with Article 8(2) of Council Directive 91/414/EEC on the placing of plant protection products on the market). COM(2001) 444 final.* Commission of the European Communities, Brussels.

European Commission (2005) Consultation: The Future of Pesticides in Europe. Available at: http://ec.europa.eu/food/consultations/pesticide_background_en.htm (accessed June 2010).

European Commission (2006a) *Proposal for a regulation of the European Parliament and of the Council concerning the placing of plant protection products on the market. COM(2006) 388 final.* Commission of the European Communities, Brussels.

European Commission (2006b) *Communication from the Commission to the Council, the European Parliament, the European Economic and Social Committee and the Committee of the Regions. A Thematic Strategy on the Sustainable Use of Pesticides. COM(2006) 372 final.* Commission of the European Communities, Brussels.

European Commission (2006c) *Proposal for a Directive of the European Parliament and of the Council establishing a framework for Community action to establish a sustainable use of pesticides. COM(2006) 373 final.* Commission of the European Communities, Brussels.

European Commission (2008) *Amended Proposal for a Regulation of the European Parliament and of the Council concerning the placement of plant protection products on the market. COM(2008) 93 final.* Commission of the European Communities, Brussels.

European Parliament (2007) Draft report by the committee responsible; PE388.326. Commission of the European Communities, Brussels.

Heilig, U. (2009) The new European PPP regulation: general improvements and perspectives for BCAs. Paper presented at the *Annual Biocontrol Industry Meeting*, Lucerne, Switzerland, 19 October.

Hokkanen, H. and Menzler-Hokkanen, I. (2007) Trade-off effects of current regulatory practice. Paper presented at the *REBECA Conference on Balanced Regulation for Biological Plant Protection*, Brussels, 20–21 September.

Hokkanen, H. and Menzler-Hokkanen, I. (2008) Deliverable 24: cost, trade-off and benefit analysis. In: *Regulation of Biological Control Agents: Final Activity Report*. Christian-Albrechts-University, Kiel, Germany.

Holm, R.E., Baron, J.J. and Kunkel, D.L. (2005a) The Ir-4 programme and its cooperation with the crop protection industry to provide new pest control solutions to US speciality crop growers. In: *The BCPC International Congress Proceedings: 2005*, Vol. 1. British Crop Protection Council, Alton, UK, pp. 239–250.

Holm, R.E., Baron, J.J. and Kunkel, D.L. (2005b) World horticulture in crisis. In: *The BCPC International Congress Proceedings: 2005*, Vol. 1. British Crop Protection Council, Alton, UK, pp. 31–40.

Hood, C., Rothenstein, H. and Baldwin, R. (2003) *The Government of Risk*. Oxford University Press, Oxford.

House of Commons (2005) *Progress on the Use of Pesticides: The Voluntary Initiative*. Environment, Food and Rural Affairs Committee, HMSO, London.

IBMA (2009) IBMA request for clarification of position of 'grey area' products in the UK. Internal document submitted to CRD. International Biocontrol Manufacturers' Association, Mourenx, France.

Larsen, H.J. (2004) The use of green taxes in Denmark for the control of the aquatic environment. Paper presented at *OECD Workshop on Evaluating Agri-Environmental Policies*, Paris, 6–8 December.

Lindsay, A.E. (2005) Factors shaping the future of the US pesticide regulatory system. In: *The BCPC International Congress Proceedings: 2005*, Vol. 1. British Crop Protection Council, Alton, UK, pp. 341–348.

PSD (2008a) *Rolling Three Year Business Plan 2008–11*. Pesticides Safety Directorate, York, UK.

PSD (2008b) *Assessment of the impact on crop protection in the UK of the 'cut-off criteria' and substitution provisions in the proposed Regulation of the European Parliament and of the Council concerning the placement of plant protection products in the market*. Pesticides Safety Directorate York, UK. Available at. http://www.pesticides.gov.uk/approvals.asp?id=1946 (accessed 13 July 2010).

RCEP (2005) *Crop Spraying and the Health of Residents and Bystanders*. Royal Commission for Environmental Pollution, London.

RCEP (2006) *Response to the Commentary of the Advisory Committee on Pesticides on the RCEP's Report on Crop Spraying and the Health of Residents and Bystanders*. Royal Commission for Environmental Pollution, London.

REBECA (2008) Regulation of Biological Control Agents: Final Activity Report. Christian-Albrechts-University, Kiel, Germany.

Richardson, D. (2005) The registration process, its effect on active substance availability, and initiatives to reduce the impact on minor crops at both UK and EU level. In: *The BCPC International Congress Proceedings: 2005*, Vol. 1. British Crop Protection Council, Alton, UK, pp. 231–238.

Waage, J.K. (1997) Biopesticides at the crossroads: IPM products or chemical clones? In: Evans, H.F. (chair) *Microbial Insecticides: Novelty or Necessity? BCPC Symposium Proceedings No. 68*. British Crop Protection Council, Farnham, UK, pp. 11–20.

6 Policy Networks, Change and Innovation[1]

Scientific research can identify sustainable benefits from biopesticides. However, the regulatory system may not function in a way that allows the potential benefits for the rural economy, e.g. the consistent supply of quality food more acceptable to customers, to be secured. As Chapter 5 has shown, biopesticide development was (until recently) locked into an inflexible and unimaginative chemical pesticide model. The system of regulation was preventing social benefits from being captured. Given, however, that existing actors in the policy network are primarily oriented to chemical solutions to pest infestation problems, how has change been brought about? Policy network theory points to policy networks being good at managing incremental change, but tending only to innovate in conditions of crisis or exogenous shock. A complicating factor is the role of the EU in pesticide legislation. Its system of decision making and in-built 'checks and balances' is not designed for rapid policy change or paradigm shifts.

This chapter considers policy actors in biopesticides and pressures for change in terms of a policy network perspective. It analyses the small size of the policy network and its relative lack of political sophistication and focuses on the notion of 'regulatory innovation'. In other words, given that regulators are risk averse, under what conditions can regulatory innovation occur? It builds on Black's framework (2005) to provide an assessment of the relative importance of endogenous and exogenous pressures for change. Our particular focus is on regulatory innovation within CRD in the area of biopesticides. By using the insights of policy network theory and interviews with key actors we can identify the change agents and processes that have created a momentum towards regulatory innovation. The question is whether environmentally friendly scientific and technological innovations in pest control on food crops can be matched by appropriate regulatory regimes that meet concerns about public safety and environmental impact, but do not unduly constrain developments that help the achievement of sustainability goals for the rural economy.

© CAB International 2010. *Biopesticides: Pest Management and Regulation*
(A. Bailey et al.)

Policy Network Analysis

Richardson and Jordan (1979) explained the way in which policy making in Britain was disaggregated into a number of subsystems giving pressure groups ample opportunities to influence policies of concern to them. They argued that the policy-making map was made up of a series of distinct vertical compartments and these tended to be organized around a government department and its client groups, largely closed off from the general public. They deliberately chose the term 'community' to reflect the intimate relationship between groups and departments.

The term 'policy network' has largely displaced that of 'policy community' in the literature. Indeed, policy network analysis has become the dominant paradigm for the study of the policy-making process in British political science (Dowding, 1995: 136). Relatively few networks 'have the internal stability and insulation from other networks typical of policy communities' (Grant, 2000: 49). Richardson's work, moreover, suggests a shift in emphasis from a policy-making world of tightly knit policy communities to a policy process more loosely organized and hence less predictable. One can distinguish between five types of network 'ranging along a continuum from highly integrated policy communities to loosely integrated issue networks' (Rhodes and Marsh, 1992: 13). At one end of the spectrum is the policy community, which is based around stability of relationships, a highly restricted membership, 'vertical interdependence based on shared service delivery responsibility' and insulation from other networks and, usually, the general public (including Parliament) (Rhodes and Marsh, 1992: 13). At the other end of the continuum, issue networks have a large number of members with a limited degree of independence. 'Stability and continuity are at a premium, and the structure tends to be atomistic' (Rhodes and Marsh, 1992: 14). The Rhodes and Marsh spectrum has been adapted by Daugbjerg (1998a) to present policy communities and issue networks as two extreme network types on a continuum. He considers the dimensions of membership, integration and institutionalization, and considers how these vary between an extreme policy community and an extreme policy network (see Daugbjerg, 1998a: 40–44).

The biopesticides policy network

The biopesticides policy network falls clearly within the loose end of the spectrum. It is relatively weakly developed compared with other networks analysed in the political science literature. Grant has commented that the 'biological control industry has the weakest policy network I have encountered' (cited in Chandler, 2007). It often lacks political sophistication in that technical knowledge is not always matched by an understanding of political processes. It is relatively immature in the sense of being at an early stage of organizational development and some of its members have problems of limited resources and capabilities. There is also a lack of trust between some of the key actors. It is more than just a loose network; it is also incomplete in that the retailers are barely integrated into it at all. This results in network underperformance, even if it is not completely dysfunctional (Greaves and Grant, 2010). The following potential participants in the biological control policy network can be identified at member state level:

1. The regulatory agency (which may often form the hub of the policy network).
2. The growers (and their representative organizations).
3. The biocontrol manufacturers (and their representative organization).
4. Consultants (who can be important intermediaries).
5. Environmental non-governmental organizations (NGOs).
6. Retailers.
7. Consumer organizations.
8. Academic researchers.

DEFRA is excluded from this list as, under governance arrangements, its role should be one of 'steering'. It is expected to be softer, less intrusive and less hierarchical than under traditional systems of government. In short, 'there has been a shift from government by a unitary state to governance by and through networks' (Rhodes, 2006: 430). Governance is seen as signifying:

> ... a change in the meaning of government, referring to a new process of governing; or a changed condition of ordered rule; or the new method by which society is governed. I employ a stipulative definition; it refers to self-organizing, interorganizational networks.
>
> (Rhodes, 1997: 35)

The Rhodes formulation makes clear that policy networks facilitate negotiation and the development of shared understandings among participants. In order to function properly, policy networks must be constitutive of all relevant stakeholders.

Network components

We need to consider each of the components in the self-steering policy network. The national regulatory agency can play a key role in both creating and sustaining a policy network. CRD can be seen as the 'hub' of the network and it has devoted considerable resources to stakeholder engagement. It has set up regular joint liaison arrangements with the IBMA and its Availability Action Plan Implementation Group comprises a range of stakeholders. Nevertheless, it is constrained in the initiatives it can take, both in terms of the existing pesticides legislation (both EU and UK) and its mandated aims and objectives. The objectives of the organization appear consistent with promoting the wider use of biopesticides, but perhaps only with further ministerial approval and guidance. At a practical level, the approvals side of CRD is set up with scientific staff to undertake the task of registration to ensure the safe use of pesticides; it may not, therefore, be equipped to take on an advocacy role. The role of the 'biopesticides champion' is to assist biologicals through the registration process, not to be an advocate in any stronger sense of the word. As a senior official within the agency put it, 'My challenge is to promote the scheme, not to promote biopesticides, there is a difference'. While there are differences of culture within CRD, the Director of Approvals has been very actively involved in network formation and development activities. The challenge had been that 'we particularly felt that we were not meeting the right stakeholders and they were not hearing us. You had to sit back and ask why they are not listening to us' (interview, 8 December 2005). One reason for this was the incomplete policy network.

A particular challenge for farmers and growers is the withdrawal of plant protection products as a result of the EU regulatory review process. The National Farmers' Union (NFU) is active within CRD's Minor Uses Network which has considered the contribution of biological products to filling gaps in availability. The NFU seems relatively well disposed towards CRD; indeed, in discussions about the EU's REACH regulations they recommended that PSD should form the core of a new chemicals agency. Through their organizations, farmers and growers are relatively well integrated into the policy network.

Despite undergoing organizational development, the IBMA has been hampered by a lack of resources and the fact its technical knowledge is not always matched by a comparable level of political sophistication. IBMA has often had difficulty in acting in a proactive fashion and portraying itself as an authoritative spokesperson for the industry that can make effective decisions about its stance on issues sufficiently quickly. It has also not organized all potential registrants of biopesticides, which is a challenge for CRD in their outreach efforts directed at the industry. In addition, previous experience with the regulatory system has, to some extent, undermined the confidence of product developers. Even when they do make contact they may be reluctant to provide relevant information, making it difficult for CRD to assist them. For those firms that do make contact, our observations have shown that pre-submission meetings are vital parts of the process.

PSD had to build up its relationship with IBMA in order to find a route into the industry. As a senior CRD official put it, '[We] had to build up confidence, [we are] now much closer to IBMA, [we] had to break into them, [we went] out there telling them there is a plan, but they were reluctant to come and meet us'. Another official summed up: 'It's a new relationship with the IBMA. They've offered us visits round plants – formulation technology, unfamiliar techniques, opportunities to see it in a field.' One practical indication of this new relationship is the joint working group of IBMA and CRD on efficacy issues. IBMA is also a member of the Availability Action Plan Group and CRD and others have been invited to IBMA meetings. The REBECA programme, funded by the European Commission, has also brought together relevant actors who might not otherwise have had contact. It has also been argued that the annual conference of the IBMA in Lucerne is emerging as a 'one-stop shop' for the policy network.

Given the relative fragmentation of the policy network, it might seem that specialist consultants would be able to play an intermediary role in bringing actors together and using their technical expertise to devise policy solutions. To some extent, consultants do play such a role. For example, they are prominent in the IBMA and often form the IBMA delegations that interact with the CRD, e.g. in the joint efficacy working group. There is, however, some ambivalence about them within the hub of the network. One perception within CRD is that 'we do consultants' job for them' in the sense that consultants ring up and ask questions that they incorporate into advice that they sell to their clients. There is concern that in certain instances they could convey the impression to some clients that access to the regulatory system is more difficult than is the case and this can produce some suspicion of their role.

Environmental NGOs tend to have a wider remit than pesticides, with the exception of the Pesticides Action Network (PAN). There is a lack of engagement by such groups in the biopesticides debate, reflecting indifference rather than hostility. For its part, IBMA was slow to reach out to environmental groups as potential allies (and still finds it difficult to establish a dialogue), which is disappointing given that their members are producing more sustainable products than conventional pesticides. Environmental groups have often been relatively isolated, with the debate about a more sustainable agriculture being framed more around discourses about organic farming. Although they have generally been critical of pesticides and called for greater restrictions on their use, environmental groups have not been particularly supportive of biological alternatives. This may be in part because of a suspicion that they are 'still pesticides'.

Retailers often push for levels of pesticide reduction more rigorous than those required by regulators, which in themselves are very stringent (see Chapter 7). Our research suggests a lack of connection between large supermarkets and the rest of the policy network. They do not tend to actively promote biopesticides to their growers as a sustainable alternative to synthetic pesticides, arguing that they cannot promote particular commercial products. Marks & Spencer and Sainsbury's are two exceptions; indeed, the latter held a conference in 2008 to discuss advances in the use of biopesticides with their suppliers. CRD sources have stated that links with retailers were relatively tenuous and this was confirmed by our interviews with large supermarket chains. One supermarket chain commented: '[We] only interact with PSD if they want specific information from us'. Supermarkets put more emphasis on using category management and their own supply chains than building relations with CRD. There is thus a lack of effective engagement between the hub of the network, CRD, and a key set of commercial actors that are pursuing their own pesticide policies (see Chapter 7).

Retailers see themselves as proxies for the consumer, and consumer organizations are not particularly involved in the discussion of biopesticides. Consumers tend to have a clear, if rather ill-informed, image of organic produce, but relatively little understanding of the potential contribution of biocontrol agents to a more environmentally sustainable agriculture. As one retailer put it when referring to the relative ignorance of consumers, 'You do get some daft responses. If you ask them about organics, they say no pesticides are required, when you explain there are pesticides applied, they get very upset.'

Academic researchers can help link various participants into a more effective network. We feel that our research has facilitated information exchange among relevant actors and heightened an awareness of the contribution of biopesticides. Our workshops, moreover, have helped bring network participants together. Academics must be careful not to cross the dividing line between being analysts and advocates. Nevertheless, they should be able to win the trust of participants whose interests or perspectives do not always coincide and hence facilitate constructive dialogue and the identification of policy solutions.

Policy networks at the EU level

Incomplete or fragmented policy networks are even more evident at EU level. There is an absence of any functional equivalent at the EU level of the national regulatory authorities which can serve as a hub around which a policy network can cluster. Such regulatory bodies can provide a location for meetings. They also have authority resources, meaning the chance of influencing the way in which those resources are used creates an incentive for meeting them. Farmers' organizations, the IBMA and environmental organizations are all present in Brussels, but the greater complexity of the decision-making process resulting from co-decision makes it even more difficult to focus representative efforts efficiently than at a national level.

However, REBECA (www.rebeca-net.de) filled a significant vacuum by creating a neutral yet informed policy space in which various actors could interact. The attendances at their conferences showed the considerable level of interest in the subject of biological control agents, but also the relative lack of opportunities to interact on a systematic basis. Regulators can also meet in the Biopesticides Steering Group of the OECD or in various EU-level committees. Informal, bilateral links between regulators are also continually developing. The IBMA annual conference provides one meeting point. However, with the end of REBECA, there is no general umbrella framework that can facilitate the discussion of issues at a European level among a wide range of actors. One consequence has been that debates in the European Parliament have not always been as well informed as one might hope.

Policy Networks and Change

Policy communities have high entry barriers around them and can become rather exclusive networks of well-established insider groups. Policy networks give structure to the decision-making process, providing outsiders and insiders with different opportunities for either changing or maintaining the existing order within a sector (Daugbjerg, 1998b: 79). Therefore, for defenders of the status quo, a sectoral policy network with a great deal of cohesion among its members is a powerful political resource (Daugbjerg, 1998b: 79). What emerges is an approximation of an elite cartel where participants collude so as to preserve the existing parameters of the policy-making process (Grant, 2000: 51). The objective of the policy-making process within such communities is often not solving real problems, but avoiding conflict, creating or maintaining stable relationships, and the avoidance of abrupt policy changes (Stringer and Richardson, 1982: 22).

Indeed, the most common and recurrent criticism of the policy network analysis is that it cannot explain change (Rhodes, 2006). Richardson (2000) summarized such arguments in a seminal article. Essentially, policy network analysis stresses how networks limit participation in the policy process, decide which issues will be included and excluded from the policy agenda, shape the behaviour of actors through the rules of the game, privilege certain interests, and substitute private government for public accountability. It is about stability, privilege and continuity. Richardson observes that the 1980s and 1990s witnessed much policy change and instability in West European states. In particular, governments adopted a more impositional policy style, and interest groups learned to exploit opportunities presented by a policy process that was increasingly characterized by multiple opportunity structures. Richardson focuses on the possible causes of policy change, including the importance of state power (e.g. government has been more assertive in attacking some of the old distributional coalitions or policy communities which were resisting policy change), changes in the behaviour of groups as they adjust to and exploit opportunities presented by multi-arena policy making, and the impact of new policy fashions, reflecting knowledge and ideas which can act as a virus-like threat to existing policy communities.

There have been three main attempts to build the analysis of change into policy networks: advocacy coalitions, the dialectical model and decentred analysis (see Rhodes, 2006: 436–438). The advocacy coalition framework (ACF) has four basic premises, helpfully summarized by Sabatier and Jenkins-Smith (1993: 16): (i) 'understanding the process of policy change ... requires a time perspective of a decade or more'; (ii) 'the most useful way to think about policy change ... is through a focus on policy subsystems'; (iii) 'those subsystems must include an intergovernmental dimension'; and (iv) 'public policies ... can be conceptualized in the same manner as belief systems, that is, sets of values priorities and causal assumptions about how to realize them'. Sabatier and Jenkins-Smith argue that coalitions attempt to translate their beliefs into public policy and their belief systems determine the direction of policy. Their resources determine their capacity to change government programmes. Resources change over time, usually in response to changes external to the subsystem. Furthermore, they distinguish between core and secondary beliefs and argue that coalitions have a consensus on their policy core resistant to change. In contrast, secondary aspects of the belief system can change rapidly (see Sabatier and Jenkins-Smith, 1993: 25–34).

Marsh and Smith (2000), meanwhile, suggest a dialectical model whereby change is a function of the interaction between the structure of the network and the agents operating in it, the network and the context in which it operates, and the network and policy outcomes. Networks are structures that can constrain or facilitate action but do not determine actions because actors interpret and negotiate contracts. Exogenous factors may prompt network change, but actors mediate that change. We must examine not only the context of change, but also structures, rules and interpersonal relationships in the network. Finally, not only do networks affect policy outcomes, but policy outcomes feed back and affect networks.

Third, there are those who propose an interpretative turn and argue that policy network analysis should make greater use of ethnographic tools such as: 'studying individual behaviour in everyday contexts; gathering data from many different sources; adopting an "unstructured" approach; focusing on one group or locale; and, in analyzing the data, stressing the "interpretation of the meanings and functions of human action"' (Rhodes, 2006; paraphrasing Hammersley, 1990). Bevir and Rhodes (2003: 62–78) argue for a decentred study of networks, for a shift of emphasis from institution to individual, and a focus on the social construction of policy networks through the ability of individuals to create meaning. Although we take a scientific realist as opposed to a social constructivist approach (see Chapter 2), elite interviewing and observations are ideal methods if 'one is interested in actors' perceptions of the world in which they live, the way in which they construct their world and the shared assumptions that shape it' (Burnham *et al.*, 2008: 246).

Rhodes (2006) concludes that 'all three approaches to network change are part of a broader trend in political science in exploring the impact of ideas on policy making'. Sabatier's work on advocacy coalitions, for example, stands alongside that of Kingdon's (1984) on policy ideas and policy agenda. Richardson (2000) discusses this link between changing policy networks, new ideas and setting policy agendas. Ideas and knowledge can, he says, 'upset the cosy life of established policy communities and networks' (Richardson, 2000: 1021). They can provide a real challenge to stakeholders who have relied on the security of cocoon-like policy communities. In practice new ideas and policy frames often 'capture' all stakeholders, who often find themselves adjusting to a new set of rules and power distributions quite different from the old policy regimes.

Of course, some of the network literature *does* engage with the issue of change. Daugbjerg (1998a), for example, develops a theoretical model which establishes a causal link between network types and policy change. He argues that where policy change is put on to the agenda, the kind of network that exists has an important influence. With a cohesive network, policy change is likely to be moderate because network members can form strong coalitions against outsiders. In contrast, radical policy change is most likely where a non-cohesive network exists. In such a network, coalition forming will be difficult (Daugbjerg, 1998a: 188–189). Greaves and Grant (2010), however, have challenged some of these theoretical propositions. Coalitions may be less likely to occur in loose networks, thereby allowing outsiders in and promoting 'network change'. However, it is unclear whether the necessary interactions for policy change can then take place. Even if a looser network can promote change in the sense of putting a plan together, it may not be able to be implemented effectively. This could be a particular problem with incomplete networks. It is likely, therefore, that policy change requires a certain degree of cohesion in the network, even if rigid policy communities will often impede change (Greaves and Grant, 2010). Greaves and Grant suggest, therefore, that changes in the biopesticides regulatory framework have occurred *despite* the loose network. It was in large part due to pressure from the Cabinet Office. The authority resources of the state were used to produce an outcome (thereby supporting Richardson's notion of state power being used to drive through change). A somewhat more cohesive network may, however, have promoted greater change; indeed, regulatory reform has coincided with some strengthening of the network in recent years (Greaves and Grant, 2010).

Marsh's dialectical model also provides some leverage, in particular in terms of feedback effects from policy. Government attempts to encourage biopesticides have resulted in regulatory reform and changing attitudes within CRD, which has fed through into improvements in integration within the network, which in turn has resulted in improved outcomes in terms of rates of registration. Marsh and Rhodes, moreover, argue that factors exogenous to policy networks lead to changes in both the network and the policy outcome. To understand and explain policy change, we need to understand and explain network change. As Marsh and Rhodes (1992: 257) put it:

> Most (network) change is explained in terms of factors exogenous to the network, although the extent and speed of the change is clearly influenced by the network's capacity to minimize the effect of such change.

Intervention from central government departments would count as exogenous pressure, given the way we have defined the biopesticides network. In Marsh and Rhodes' view, however, the driver for change in networks and outcomes lies in broader economic and political change and changes in knowledge. Certainly, there are contextual drivers suggesting that biopesticides may be an 'idea whose time has come'. First, the public is concerned about the possible health effects of pesticide residues on food. Second, there is consumer preference, even when there are no safety concerns, for a reduction in pesticide residues in food, leading to action by retailers, consumers and the Food Standards Agency (FSA). There is also a requirement to integrate chemical pesticides with alternative methods in order to develop systems of crop protection which are sustainable. This has an ecological dimension; crops need to be protected using methods which do not damage the environment, in particular in terms of water pollution and biodiversity. Public concern over the impact of pesticides on the environment, therefore, is a third driver of change. Pesticides are also required that prevent the development of resistance by the pest to the control agent. Following the implementation of European Directive 91/414/EEC there has been a significant decline in the number of active ingredients permitted for use in crop protection products. Moreover, because of the expense of research and registration, manufacturers are unlikely to develop new chemical products on a large scale. As there has been a reduction in the number of pesticide products available for use, this increases the problem of pesticide resistance. The broad solution is to use IPM, as outlined in Chapter 1.

These contextual drivers may, at the margins, have promoted network change, or at least a greater integration of actors into the network. They are similar, certainly, to the exogenous pressures referred to by Marsh and Rhodes. However, we should not overstate their importance in pushing through change. The public may be concerned about pesticide residues but are likely to have no clear view on biological alternatives, of which they are not well informed. They may also be put off by the term 'pesticides'. Retailers, as we have seen, have done little to engage in the debate. Although growers are particularly affected by problems of resistance and the withdrawal of plant protection products, they are 'policy takers' rather than 'policy makers'. They operate within the constraints of a stringent regulatory framework and have to cope with the market power of the supermarkets. Linking back to the literature on the influence of ideas on policy networks, biopesticides are not taken adequate account of in debates on sustainability. This debate tends to be polarized between conventional and organic alternatives with insufficient attention paid to IPM. Biopesticides do not have a particularly high profile among decision makers.

There are two final issues relating to policy networks relevant to this discussion. First, the policy network literature tends to refer to change as opposed to innovation (see Chapter 1 for a distinction between these two terms). Arguably, it should make more use of the latter. Second, policy network analysis often fails to discuss what it means by 'policy'. Indeed, the notion of policy or policy outcomes does not provide the full picture in an age of governance. The modern meaning of the English notion 'policy' is that of a course of action or plan, a set of political purposes – as opposed to 'administration' (Wilson, 1887). Above all, as Parsons suggests, we can see policy as a rationale, a manifestation of considered judgement. It is an attempt to define and structure a rational basis for action or inaction (Parsons, 1995: 14). Policy can be seen as culminating in a government decision. Birch (1979; cited in Jones, 2001) distinguished between two broad kinds of government decisions: (i) rules, regulations and public pronouncements; and (ii) public expenditure and its distribution. Does such an approach need updating, however, to reflect a new world of governance and the regulatory state (see Moran, 2005)? As outlined in Chapter 1, in the regulatory state many decisions are taken by government agencies.

However, can decisions taken by such agencies be defined as policy? Osborne and Gaebler (1992) make the distinction between policy decisions (steering) and service delivery (rowing). They argue that bureaucracy is a bankrupt tool for rowing, and in its place they propose entrepreneurial government, with its stress on working with the private sector and responsiveness to customers. This transformation of the public sector involves 'less government' or less rowing but 'more governance' or more steering. In essence, this is a definitional issue; it does not alter the fact that policy networks could impact on government agencies in the same way they may impact on traditional policy outcomes. Furthermore, CRD has a policy advisory role and some of its activities (such as the Biopesticides Scheme) go beyond routine matters and can be defined as policy.

Having said this, the risk averseness of regulators means it may be difficult to drive through the necessary change. Moreover, we are left with the problem of whether network change can feed into significant policy change if the overall network remains very loose and incomplete. It is our view that the incomplete network helps to explain why change has not been as significant as it could have been. However, as we have seen, policy network analysis on its own cannot fully account for the change that has taken place. We propose, therefore, to move towards policy network analysis and focus on exogenous pressure (from government departments) and endogenous pressures (from within CRD).

Regulatory Innovation

Innovation remains a central challenge for systems of regulation. As Greaves (2009: 245) notes, 'Bureaucrats and regulators are typically risk averse. The desire to avoid things going wrong means they are not natural innovators. Risk averseness does not create an encouraging environment for regulatory innovation (indeed, the term is almost a contradiction).' As we will see, regulatory innovation has occurred within CRD through an interaction of endogenous and exogenous factors. These have come together to create a window of opportunity through which regulatory innovation has been able to occur.

Regulatory change, innovation or adaptation?

Does, however, the change count as innovation? To return to Hall's typology (as outlined in Chapter 1), some aspects of the Biopesticides Scheme are essentially first-order policy changes (e.g. lower fees for biologicals). Most, however, fall within the category of second-order policy changes (e.g. a biopesticides champion, pre-submission meetings, etc.). In some ways, lower fees are the biggest change but, as Black suggests, first-order changes may be significant in terms of scale or impact whilst not strictly counting as innovation (Black, 2005a). Taking everything together, we can describe what has occurred as regulatory innovation; to return to Black, it has involved 'new solutions' (Black, 2005a). Regulatory adaptation, on the other hand, would imply more of a modification or alteration of existing practice, or incremental change as opposed to the more radical reform which has taken place. It is an unusual step for a regulatory agency that usually has to stick closely to what is laid down in statute to negotiate new policy spaces in which to operate (which is essentially what has happened here). In other words, PSD sought clearance from DEFRA, particularly in relation to the funding of the scheme, thereby creating a new space in which to take action. This has extended their formal remit, albeit cleared by ministers. As their Director of Approvals puts it, this has been 'quite remarkable for a regulatory agency' (Sainsbury's Conference, 18 March 2008).

Black's five worlds

We introduced Black's five worlds of regulatory innovation in Chapter 1. It will be helpful now to elaborate on them a little. An emphasis on champions is found in much of the work in public sector innovation. The characteristics of such champions are risk preferring, being open to new ideas, persuasive, empathetic and occupying key strategic positions within organizations. Generally speaking, such individuals will have sufficient financial resources to absorb losses from unprofitable innovations, be able to cope with a high degree of uncertainty, and have a favourable attitude towards change and risk taking. In the world of the individual, innovation is explained by one, or sometimes two, key individuals who are able to push their performed innovation through critical decision junctures or 'policy' windows. This relates to the work of Kingdon (1984) who argued that 'policy entrepreneurs' take advantage of policy windows offered by the concatenation of policy problems, policies and politics to catapult new items on to the political agenda and change policy.

The organizational world suggests that innovation is fostered by organizations in which: the leader or leaders have a positive attitude towards change; there is low centralization; its members have a high degree of knowledge and expertise; procedures are not highly formalized; there is a high degree of interpersonal connections between the organization's members; and there is a high degree of organizational slack, for example, the extent to which uncommitted resources are available to the organization, with the biggest factor being size – *ceteris paribus*, large organizations are more innovative than smaller ones. Also important are the existence of cultures that are supportive of risk taking and can tolerate mistakes and failures.

The work on innovation within government ('the state world') comes mainly from political science. As Black writes, 'if the question "how and why does innovation in public policy occur" is rephrased as "how and why does policy change occur" or "how and why does policy learning occur", then immediately almost any theory of public policy formation would have an answer' (Black, 2005b: 25). Public choice theory would state that innovations are political goods which are sold to the highest bidder or coalition of bidders, and will depend on the distribution of costs and benefits of particular groups. Pluralism would argue that they come about through an interplay of interest group pressures. Public opinion response theory would state that they are the result of public pressure, mediated by the media. Rational theories of bureaucracy would argue that they are the result of self-interested bureaucratic decisions (Downs, 1967).

It is helpful to focus on such rational choice theories. Bureaucracies have to cope with considerable forces of inertia, as outlined in the classic text *Inside Bureaucracy* by Downs (1967: 195–197). First, like most organizations, bureaucracies have a powerful tendency to continue doing today what they did yesterday. This is because established processes represent an enormous previous investment in terms of money, time and effort. If new behaviour patterns are adopted, these costs must be faced again.

Downs believes that the more officials that are affected, the greater resistance will be to significant change. Therefore, the larger an organization is, the more reluctant it will be to change, and small bureaus tend to be more flexible and innovation minded than larger ones. Second, self-interest motivates officials to oppose changes which would result in net reductions in things they personally value, such as personal power, prestige and income. Therefore, officials will tend to oppose changes that would lead to a net reduction in the amount of resources under their own control; and changes that would decrease the number, scope or relative importance of the social functions entrusted to them. Downs explains that this is why transfers of functions from one section to another are often resisted by the sections losing functions.

Downs, however, suggests some drivers for change (Downs, 1967: 198–200). First, there is the desire to do a good job. This could be due to loyalty to specific parts of the bureaucracy, to specific ideas or to society as a whole. This motive will be particularly prevalent in the creation of new bureaus or new sections within an existing bureau. Second, there is the desire for aggrandisement. Downs writes that self-interest is a powerful cause of inertia, but it can motivate change if officials receive greater rewards for altering the status quo than preserving it. 'The greatest of such rewards are gains in power, income and prestige associated with increases in the resources controlled by a given official or a given bureau' (Downs, 1967: 198). Politicians, however, are more reluctant than officials to increase the total size of the government budget. Officials, therefore, have a better chance of getting their resource-expanding innovations improved if they can reduce expenditures elsewhere. Therefore, proposed innovations must carry out social functions performed elsewhere, leading officials with a powerful motive to 'capture' functions performed by other bureaucracies. A third motive for change in bureaus is self-defence against pressure from external agents, such as abolition or threats to reduce its resources. Bureaucracy threatened with abolition, for example, must find new functions or reinstate the importance of its present ones. To some extent Down's analysis was 'undermined by manifest evidence that bureaucrats satisfied rather than maximised' (Deakin and Parry, 2000: 63). Dunleavy's (1991) bureau shaping model came to be influential, suggesting that 'size and aggrandisement were not ends in themselves' but that 'bureaucrats would calculatedly "shape" their task to obtain the optimal mixture of rewarding and achievable work' (Deakin and Parry, 2000: 63).

There is not always such a clear distinction between change and innovation in the literature as Black (2005a) suggests: Downs (1967), for example, seems to use the terms interchangeably. That being said, the literature in the state world directly associated with 'innovation' is rather narrow. The institutional literature, however, has focused to some extent on innovations, and provides a broad set of arguments. Central to 'new institutionalism' is that 'institutions matter' as they provide the structure to which the action and interaction occurs (March and Olsen, 1984). The approach suggests that innovation is explained by the impact of institutional structures on decision making by political actors, including bureaucrats and those in regulatory agencies.

The 'global world' focuses on policy making by international bodies and networks. States have been relatively passive or bypassed in many areas of policy development which are nevertheless regulatory in character. Either they are willing to be instructed by international organizations or epistemic communities as to what action to take,[2] or they are bypassed by transnational organizations that set technical or professional standards (Boli and Thomas, 1997). Finally, we have 'the world of the individual'. An idea is likely to be adopted and enacted primarily on the shape and form of the innovation itself and, for some, on the extent to which it 'fits' with the prevailing cognitive/normative frameworks. 'The more the innovation is simply expressed, trialable and observable, the more it is expressed as a general theory, and the more it fits, or can be represented as fitting, with dominant cognitive and normative schemas, the more likely it is the idea or innovation will be adopted' (Black, 2005b: 40).

Through our research, we have developed an understanding of how regulatory innovation within CRD is likely to have occurred. In terms of Black's five worlds, key individuals have been vital in driving the process forward, within both the Cabinet Office and CRD. Similarly, organizational characteristics have played a role (for example, the relatively small size of the organization, a desire for knowledge among the scientific regulators, etc.). Moreover, there has been the intervention of the Cabinet Office and the impact of institutions ('the state world'). Although the 'global world' is of less significance, CRD operates within OECD guidelines, has engaged with REBECA, and the review of Directive 91/414 is clearly important. In terms of the 'world of the innovation', biopesticides fit into their surrounding environment, not least in terms of issues surrounding sustainability, pesticide resistance and the limited number of products. They are not, however, an 'idea' easy to get to grips with or widely understood.

Black can be criticized for trying to cover every possible theory or explanation: there is not much differentiation or selection, and therefore potentially little leverage. To some extent she acknowledges this, stating 'one is inevitably prey to the criticism that what is offered is simply a bewildering variety of perspectives or explanations with no clear direction as to which course one should take' (Black, 2005b: 41). Black states that her aim is to 'provide an analytical framework for much richer explorations' (Black, 2005b: 41). It is concluded that 'neither the occurrence nor the outcomes of innovation can be controlled and predicted'. 'Innovation simply cannot be engineered' (Black and Lodge, 2005: 194). We propose a framework that focuses on the contextual and exogenous and endogenous drivers for regulatory innovation (see Greaves, 2009 for a more detailed justification for this). While overlapping with Black's analysis, it provides a clearer framework for our purposes. Black's analysis may not always provide the necessary clarity. For example, individuals may be part of the world of the individual or the world of the state. There is also considerable overlap between aspects of the 'individual' and 'organizational world': it is not always clear what should be placed in which category. Similarly, there is overlap between the organizational world and much of the institutional literature. We believe that our framework will help to provide greater clarity.

Exogenous and Endogenous Pressures

The intervention of the executive is an example of exogenous pressure. The then Business Regulation Team (BRT) of the Regulatory Impact Unit of the Cabinet Office noted in 2003 that, 'although DEFRA had been funding the research and development of "alternatives" to synthetic pesticides, none had been able to obtain the authorisation required for such products to be placed for sale in the UK as plant protection products'. They argued that PSD's testing requirements 'were evidently designed to cope with standard, mass-produced synthetic chemical pesticides which, by their nature, tend to deliver very high efficacy rates, and not with this group of safer alternatives ... this appeared to be an interesting example of regulation-inspired market failure' (BRT, 2003: 19).

DEFRA hoped to encourage the wider use of biopesticides in order to achieve its sustainability objectives. However, given such limited progress the institutions of the core executive were required to intervene. In the coded language of the civil service, 'the BRT approached PSD seeking to help to establish a workable solution to this problem.' In Grant's words, 'they used their authority to lean on PSD' (Grant, 2005: 15). The fact that the Government leaned on PSD was confirmed both by a senior figure within PSD and by an industrial executive seconded to BRT to work on biopesticides. The Director of Approvals commented that 'there was a political driver but it wasn't DEFRA or growers, it was the Cabinet Office'. Furthermore, 'it was someone on secondment to the Cabinet Office, not a career civil servant' (Biopesticides Workshop, 31 October 2007). He added, 'We did need some pressure to introduce the scheme. He gave us a kick in the teeth' (REBECA Conference, 20–21 September 2007). PSD's 'aims and objectives', agreed with ministers in spring 2003, included reducing the 'negative impact of pesticides by encouraging reductions in their use, taking account of good practice, and developing and introducing alternative control measures' (PSD, 2004: 9).

One retailer put it to us: 'PSD is under pressure to try and help their policies to adapt'. The Regulatory Affairs Officer for a manufacturer and supplier of pesticides (including biopesticides) provided an interesting insight into how PSD is beginning to change. 'They want to be seen to be doing something ... [it is] now in their best interests to look at solutions. If they're the first regulatory authority to get something in place and have a way to get products to the market, other European countries will follow what they have done.' The decision to make the pilot scheme formal, for example, was almost done from a 'PR point of view'. They were told they had to do something about it. Similarly, 'someone on high said, you will have a biopesticides scheme'. Such exogenous pressure, however, needed to combine with endogenous forces; or in other words a positive response from within the regulatory body. It was reported that PSD 'was keen to discuss ways in which the pursuit of this new aim could be promoted' (BRT, 2003: 19). In other words, they realized they had to do something. In the words of Pendlington and Dickinson (2003: 23):

> The message from the Business Regulation Team's work on biopesticides is quite simple: the Principles of Good Regulation can be used as a framework for dialogue between regulator and regulated. … The successful conclusion of this project reflects well on all concerned. On the side of the regulator, there was an obvious recognition of the existence of a problem and a willingness to think creatively about possible solutions.

The Director of Policy and the Director of Approvals made the joint decision to have a pilot scheme. The latter decided to make the Biopesticides Scheme permanent, confirming that key individuals were vital in moving the process forward (Biopesticides Workshop, 31 October 2007). The Director of Approvals has stressed the importance of an effective and strong team in driving through change (Biopesticides Workshop, 31 October 2007). One consequence of the Biopesticides Scheme has been the development of an informal internal network of staff with interest and expertise in issues related to biological control (W. Grant, unpublished). These include the appointment of specialist bio-contacts. CRD staff members are used to team working as they work in groups on approval processes and the agency has a relatively informal working style and horizontal structure in which relationships are based on collegiality and mutual respect (W. Grant, unpublished). They have received training to help them to develop their skills and have been very receptive to this career development opportunity (W. Grant, unpublished). We have noticed how those working on biopesticides show great enthusiasm for their work, perhaps because of a desire to do a 'better job', or to gain new skills or expertise, or that doing the job well is a successful career-building strategy. This links back to some of the arguments of Downs (1967) and Dunleavy (1991). A senior official adds that they were 'lucky in the people they had picked to work on biopesticides, if others had been chosen it may not have worked so well' (unstructured discussion, 31 October 2007).

In one of our observations the approvals process was started by three resource managers with an initiation meeting. This, along with the approval process integrating a number of different specialisms (a large number of individuals were brought into pre-submission meetings), shows that it is a relatively horizontal process. This may be more favourable to innovation as individuals have an opportunity to learn from different specialists. Moreover, those within approvals see themselves as scientists first and as regulators second: in other words, 'scientific regulators'. They are keen to extend their scientific expertise and have shown an interest in learning about biological alternatives. This knowledge is driving regulatory innovation forward (linking back to Black, 2005b: 20), with pre-submission meeting in particular allowing CRD to build up its expertise and develop a more appropriate registration process. As the Director of Approvals puts it, 'they have built a team of specialists who understand the issues and are continuing to learn' (Sainsbury's Conference, 18 March 2008).

Institutional Reform

As a new institutionalist would claim, the structure of institutions influences the behaviour of those within them. Stakeholders have held differing views on how the regulatory structure regarding pesticides could be reformed. One suggestion in our interviews was to enlarge the role of PSD so that it became a more general chemical regulatory and inspection authority with functions from other government agencies combined into it. Another idea was that PSD should be subsumed into the Environment Agency. This, however, could change the focus of PSD work to environmental impact, which, when it comes to biopesticides, could be even more costly than demonstrating efficacy. A further suggestion was for a separate agency for biological controls. As Grant puts it, however, 'there might be insufficient work for such an agency on a country basis, but there might be a role in this area for the European Food Safety Agency at EU level' (Grant, 2005: 18).

The UK RCEP, meanwhile, published a report in September 2005 entitled *Crop Spraying and the Health of Residents and Bystanders*. This received considerable media attention and recommended that stricter controls be placed on the spraying of pesticides on crops, given that they may be causing ill health to bystanders and those living near sprayed fields. The report recommended that responsibility for pesticides policy be separate from that of the approval of pesticides. One suggestion was to move the policy function from the PSD to a unit within the Environment Directorate of DEFRA. The report added that the remaining functions concerned with the approval of pesticides could be transferred to the Environment Agency in order to ensure better coordination with wider environmental objectives. The ACP published their response to the report in December 2005. While there were parts of it with which they agreed (ACP, 2005: 3), they were generally very critical. With regard to the governance issues, they believed it to be a complex issue, and did not believe that the limited discussion in the report provided an adequate basis on which to form a judgement. They went on:

> We can see that there are advantages in close liaison between the policy and regulatory functions ... and we would like to see the case for separation in more detail. We would, however, emphasise the need, when making decisions in this area, to take into account the current excellence of the scientific and technical staff at PSD. In our view, their performance is as strong as that of the best government departments and agencies, and ahead of the majority. It also compares favourably with that of pesticide regulatory agencies in other European countries. We believe it would be most unfortunate if a reorganisation caused this valuable concentration or expertise to be lost.
> (ACP, 2005: 33)

Of course, there is a tendency for governments to bring about institutional changes in response to criticism as a substitute for making changes in policy. For example, if PSD was incorporated into the Environment Agency, would its functioning change that much, particularly if the same people continued to work in the same distinct office in York? That being said, one particular way of encouraging innovation may be through systematic reviews of provision. The Hampton Review on UK regulation, set up by the Chancellor of the Exchequer, reported in April 2005 and proposed streamlining the regulatory structure on the grounds that there were too many small regulators (Hampton, 2005). A Better Regulation Executive was set up in the Cabinet Office to deliver the reforms. As Grant puts it, 'what is evident is that the core executive, in terms of both the Cabinet Office and the Treasury, are actively engaged in issues of regulatory structure' (Grant, 2005: 18). Indeed, the Hampton Review was prompted in part by the Treasury who are concerned about the cost and powers of regulators. Part of the broader political context, moreover, is the growing criticism of the government for excessive regulation.

The Hampton Report proposed streamlining the regulatory structure on the grounds that there were too many small regulators. It did not make a specific recommendation regarding PSD, which was on the cusp as far as size was concerned. However, the implication could be drawn that it should be merged into a larger more thematic regulator. DEFRA launched a formal consultation on merging PSD with the HSE, linking to the latter's existing responsibilities for biocides and REACH. Ministers decided the organizations should be merged from 1 April 2008. As part of this decision it was also agreed that strategic policy for pesticides should remain with DEFRA and operational policy should transfer with PSD to HSE (PSD, 2008: 4). There is a risk that a greater focus on chemicals strategy within the merged organization might lead to less attention being placed on biological alternatives.

Despite the thinking behind Hampton, a relatively small organization may be more flexible and responsive and better able to develop an organizational culture favouring innovative responses to new challenges. Furthermore, a body with a clear and specific purpose may be more conducive to regulatory innovation than a larger and potentially more unwieldy one. Downs (1967), as we have seen, argued that change is easier to drive through in smaller organizations: in other words, the more individuals affected the more difficult it will be to push it through. Of course, the literature on organizations suggests it is easier in larger organizations due to uncommitted resources and organizational slack. In our view, staff will be more 'accountable' for their actions in smaller organizations; in other words it is easier to note who is performing their jobs effectively. Furthermore, good working relationships and a high degree of interpersonal connections exist within the approvals side of CRD (linking back to Black, 2005b: 20); its 'close-knit' nature fostered by its size. On balance, therefore, the relatively small size of the organization and its clear purpose are likely to have been conducive to regulatory innovation.

Conclusions

Modern regulatory philosophy emphasizes a move away from 'command and control' regulation. To put it another way, there has been a move away from 'top-down', hierarchical forms of government towards more horizontal forms of governance involving negotiation between various actors. Stakeholder engagement is a key component of this new mode of governing. Policy network analysis is of relevance here and it is important to pay attention to problems arising from incomplete or fragmented policy networks. Regulatory innovation is needed to facilitate the wider availability of biological control agents, but is not easy to achieve. Reconfiguring the regulatory process involves a learning process for the regulators and a willingness to engage effectively with a range of stakeholders. Again, this links back to notions of governance and policy network analysis. The Biopesticides Scheme has not overcome all problems as the level of applications is still relatively low. This reflects in part the fragmented character of the policy network and the fact that not all developers are IBMA members.

What is generally agreed is that there needs to be more stakeholder involvement in policy design and implementation. The weak policy network in the UK is in contrast to the USA where the 'biopesticides industry as a whole pushed' for action on biopesticides, rather than the government (interview with EPA, 18 November 2005). It also contrasts with the Netherlands where there is a much more effective relationship between different stakeholders, in part through the medium of the Genoeg project. There are specific features that have made this possible, but this does not mean that there could not be more effective dialogue in the UK. Put simply, the biopesticides network in the UK provides a good and understudied example of 'network underperformance' (Greaves and Grant, 2010). In the following chapter we consider the notion of 'retail governance'. As we have seen, the sphere of consumption is insufficiently articulated in the network. This leads to a creation of a supplementary private system of regulation in an attempt to meet consumer concerns about pesticides. However, this supplementary system of regulation does little to promote biopesticides as a safer alternative.

Note

[1] This chapter draws on Greaves (2009) and Greaves and Grant (2010).
[2] Epistemic communities are knowledge-based communities with an authoritative claim to policy-relevant knowledge within their domain of expertise.

References

ACP (2005) *Crop Spraying and the Health of Residents and Bystanders: a Commentary on the Report Published by the Royal Commission on Environmental Pollution in September 2005*. Advisory Committee on Pesticides, York, UK.
Bevir, M. and Rhodes, R.A.W. (2003) *Interpreting British Governance*. Routledge, London.
Black, J. (2005a) What is regulatory innovation? In: Black, J., Lodge, M. and Thatcher, M. (eds) *Regulatory Innovation*. Edward Elgar, Cheltenham, UK, pp. 1–15.

Black, J. (2005b) Tomorrow's world: frameworks for understanding regulatory innovation. In: Black, J., Lodge, M. and Thatcher, M. (eds) *Regulatory Innovation*. Edward Elgar, Cheltenham, UK, pp. 16–44.

Black, J. and Lodge, M. (2005) Conclusions. In: Black, J., Lodge, M. and Thatcher, M. (eds) *Regulatory Innovation*. Edward Elgar, Cheltenham, UK, pp. 181–197.

Boli, J. and Thomas, G.M. (1997) World culture in the world polity: a century of international non-governmental organization. *American Sociological Review* 62, 171–190.

BRT (2003) *Regulatory Impact Unit: 2003 Report*. Cabinet Office, London.

Burnham, P., Gilland Lutz, K., Grant, W. and Layton-Henry, Z. (2008) *Research Methods in Politics*, 2nd edn. Palgrave Macmillan, Basingstoke, UK.

Chandler, D. (2007) Biopesticides in the UK: can we get regulatory innovation? Paper presented at the *40th Annual Meeting of the Society for Invertebrate Pathology*, Quebec, Canada, 20 August 2007. Available at: http://www2.warwick.ac.uk/fac/soc/pais/biopesticides/papers/ (accessed 14 July 2010).

Daugbjerg, C. (1998a) *Policy Networks Under Pressure: Pollution Control, Policy Reform and the Power of Farmers*. Ashgate, Aldershot, UK.

Daugbjerg, C. (1998b) Similar problems, different policies: policy networks and environmental policy. In: Marsh, D. (ed.) *Comparing Policy Networks*. Open University Press, Buckingham, UK, pp. 75–89.

Deakin, N. and Parry, R. (2000) *The Treasury and Social Policy*. Palgrave Macmillan, Basingstoke, UK.

Dowding, K. (1995) Model or metaphor? A critical review of the policy network approach. *Political Studies* 43, 136–158.

Downs, A. (1967) *Inside Bureaucracy*. Little, Brown and Company, Boston, Massachusetts.

Dunleavy, P. (1991) *Democracy, Bureaucracy and Public Choice: Economic Explanations in Political Science*. Harvester, New York/London.

Grant, W. (2000) *Pressure Groups and British Politics*. Palgrave Macmillan, Basingstoke, UK.

Grant, W. (2005) The challenges of interdisciplinary environmental research: the case of biopesticides. Paper presented at the *Northeastern Political Science Association Conference*, Philadelphia, Pennsylvania, November 2005. Available at http://www2.warwick.ac.uk/fac/soc/pais/biopesticides/papers/ (accessed 14 July 2010).

Greaves, J. (2009) Biopesticides, regulatory innovation and the regulatory state. *Public Policy and Administration* 24, 245–264.

Greaves, J. and Grant, W. (2010) Underperforming policy networks: the biopesticides network in the UK. *British Politics* 5, 14–40.

Hampton, P. (2005) *Reducing Administrative Burdens: Effective Inspection and Enforcement*. HM Treasury, London.

Jones, B. (2001) The policy making process. In: Jones, B., Kavanagh, D., Moran, M. and Norton, P. (eds) *Politics UK*, 4th edn. Longman, Harlow, UK, pp. 527–542.

Kingdon, J.W. (1984) *Agendas, Alternatives and Public Policies*. Little, Brown and Company, Boston, Massachusetts.

March, J.G. and Olsen, J.P. (1984) The new institutionalism: organizational factors in political life. *American Political Science Review* 78, 734–749.

Marsh, D. and Rhodes, R.A.W. (1992) Policy communities and issue networks: beyond typology. In: Marsh, D. and Rhodes, R.A.W. (eds) *Policy Networks in British Government*. Clarendon Press, Oxford, UK, pp. 249–268.

Marsh, D. and Smith, M. (2000) Understanding policy networks: towards a dialectical approach. *Political Studies* 48, 4–21.

Moran, M. (2005) *Politics and Governance in the UK*. Palgrave Macmillan, Basingstoke, UK.

Osborne, D. and Gaebler, T. (1992) *Reinventing Government*. Plume, New York.

Parsons, W. (1995) *Public Policy: an Introduction to the Theory and Practice of Policy Analysis*. Edward Elgar, Cheltenham, UK.

Pendlington, D. and Dickinson, S. (2003) New support for biopesticides in the UK. *Pesticides News* no. 61, September, pp. 22–23.

PSD (2004) *Pesticides Safety Directorate Annual Report and Accounts 2003/04*. HMSO, London.

PSD (2008) *Pesticides Safety Directorate Annual Report and Accounts 2007/08*. HMSO, London.

RCEP (2005) *Crop Spraying and the Health of Residents and Bystanders*. Royal Commission on Environmental Pollution, London.

Rhodes, R.A.W. (1997) *Understanding Governance: Policy Networks, Governance, Reflexivity and Accountability*. Open University Press, Buckingham, UK.

Rhodes, R.A.W. (2006) Policy network analysis. In: Moran, M., Rein, M. and Goodin, R.E. (eds) *The Oxford Handbook of Public Policy*. Oxford University Press, Oxford, pp. 425–446.

Rhodes, R.A.W. and Marsh, D. (1992) Policy networks in British politics: a critique of existing approaches. In: Marsh, D. and Rhodes, R.A.W. (eds) *Policy Networks in British Government*. Clarendon Press, Oxford, UK, pp. 1–26.

Richardson, J. (2000) Government, interest groups and policy change. *Political Studies* 48, 1006–1025.

Richardson, J. and Jordan, G. (1979) *Governing Under Pressure*. Martin Robertson, Oxford, UK.

Sabatier, P. and Jenkins-Smith, H.C. (1993) *Policy Change and Learning: an Advocacy Coalition Approach*. Westview Press, Boulder, Colorado.

Stringer, J. and Richardson, J. (1982) Policy stability and policy change: industrial training 1964/82. *Public Administration Bulletin* no. 39, 22–39.

Wilson, W. (1887) The study of administration. *Political Science Quarterly* 2, 197–222.

7 Retail Governance

Retailers are capable of playing an important role in the dissemination of biocontrol products. They enjoy high levels of trust from their customers, who are concerned about residues from synthetic pesticides on fresh produce. By advocating the wider use of biocontrol products by growers, retailers can do much to stimulate their adoption. However, this needs to be done in a way that does not make the work of growers more difficult. It also has to be remembered that retailers are in business to make a profit and there are some tensions between their private systems of regulation and the state-backed system. This chapter draws on interviews with retail managers and others; the grammar and sense of the quotations have been improved.

Retailers have been permitted to assume functions that might otherwise be regarded as public in character, extending the scope of private retail governance. This can set up tensions with the public world of regulation, often referred to as the regulatory state (see Chapter 1), as it can mean two systems of 'regulation' operating alongside each other. In a strict sense, private regulatory governance is not 'regulation' in that it involves the issue of authoritative decisions by a state body to private actors. Instead, it relies on the use of contractual mechanisms to achieve desired goals. The market power of large retailers means that standards that they impose are in practice mandatory. One regulatory agency respondent argued in interview: 'Growers are concerned that more restrictions are being put on how they grow produce and this is giving an advantage to imported produce. They see this as regulation, which it isn't, it is guidance.' However, the effect on the private actor is broadly the same, it is just that the potential penalties are different: in one case, a fine or similar penalty; in the other, the withdrawal of an important contract.

The chapter draws in particular on interviews with major supermarkets. Senior technical managers (all in the fresh produce area) at head office were generally the respondents in supermarket chains, although two different managers were interviewed in the largest retailer and one other retailer sent three managers and two consultants to be interviewed at a university location. Because of considerations of commercial confidentiality, retailers are identified by letters A to H.

The Economic Power of Retailers

Retailers play a key role in British political economy, but one that is relatively understudied. Britain has a highly concentrated food retail sector in particular with one firm, Tesco, accounting for some 30% of total grocery sales. Along with three other companies (Asda, Sainsbury's and Morrisons) it accounts for 76.5% of the UK grocery market (Food Ethics Council, 2005: 5). Retailers of food and household goods play a key role in determining rates of inflation which are an issue of crucial concern to government, even if interest rate setting has been depoliticized. In August 2007, for example, the Consumer Price Index fell sharply, in part because of price cutting by retailers. However, retailers have a wider role in relation to other issues of public concern such as food safety and food quality.

The post-war period has seen a shift of power down the food chain, not just in the UK, from the farmer and the food processor to the retailer. In the first three decades after World War II, producers were encouraged and subsidized to maximize production against the background of the experience of food scarcity at the end of the armed conflict in Europe. There were also fears about food security in the context of the Cold War. Over time concerns developed about the budgetary costs of the subsidies paid to farmers, their distorting impact on world trade and the consequences that intensive farming had for animal welfare and the environment. Nevertheless, farmers still receive substantial subsidies in almost all OECD countries, the principal exceptions being Australia and New Zealand.

The retail sector was still very fragmented in the UK and elsewhere in the decades after World War II, even though there were a number of multiple chains. Over time the UK grocery market came to be dominated by a small number of retailers. Although it has happened more slowly there, this pattern is being replicated in continental Europe. The factors that have favoured concentration at a national level have included improvements in road transport and air freight which have facilitated the development of centralized depots that then service individual stores. Products can be sourced from across the world in a flexible way that requires the producer to adjust to variations in demand resulting from factors such as changes in the weather. The development of information technology has allowed much more precise control of stock ordering systems. This in turn permits more efficient supply chain management as well as the recording and analysis of precise data, related to demographics, about consumer purchasing patterns.

As the buying power of the big retailer groups has become more concentrated, one has seen a second phase of food regulation 'dominated by supply chain management and food standards strategies, designed and increasingly applied by the large multiple food retailers' (Marsden *et al.*, 2009: 124). Rather than the state having a key role, 'food regulation was driven primarily by the way food safety issues are perceived by large food retailers; leaving the State mainly to act as auditors rather than enforcers of the mainstream process' (Marsden *et al.*, 2009: 125). Using their market power the large multiple retailers have been able to impose increasingly demanding conditions on manufacturers and primary producers. Manufacturers may be required to pay for the privilege of their goods being prominently displayed in the store, asked to discount products in two-for-one or similar offers or to contribute to the cost of new store openings. Primary producers face increasingly demanding contracts in terms of price, quality and food safety standards, backed up by regular inspection visits in accordance with the slogan 'if you don't comply, you can't supply'. It can be argued that 'the oligopsony power of European retailers has been strengthened by producer-led food safety and quality assurance schemes' (Food Ethics Council, 2005: 5).

Marsden *et al.* (2009) identify the development of a third hybrid public–private phase of food regulation involving greater institutionalization at both the domestic and EU levels. This represents 'a complex model of good governance in which public and private interests are bound together in ever more sophisticated ways to deliver safe food' (Marsden *et al.*, 2009: 204). They see a blurring of the public and private in relation to food safety as particularly apparent in relation to risk because private and public actors have complementary interests in managing food risk to maintain public confidence in the foods they consume. However, they do not see these developments as diminishing the role of retailer-led private regulation. Retailer power is not diminished even though there may be some reconfiguration of its relationship with the state.

The Political Power of Retailers

The growing economic power of retailers has been matched by a growth in their political power, reflected in the formation of pressure groups that target retailers. This growth in political power has not been through the conventional route of trade associations or even through direct contacts between major companies and politicians, although these are not without importance. However, the main emphasis is on the development of a private system of governance by retailers that meets consumer concerns and goes beyond what is required by government-designed systems of regulation and also assists in the delivery of government policy objectives. As Marsden *et al.* (2000: 28) explain:

> By the early 1990s in Britain … there was a discernible switch in public policy to shift responsibility for food matters to the retailers; the major retailers have clearly become significant actors in the promotion and implementation of, for example, health policy … Reciprocally, the regulatory state has become critically dependent upon the continued economic dominance of the retailers in their role as the major provider of quality food goods.

Marsden *et al.* (2000) argue that, 'As a result, in a very real sense, [the retailers] act on behalf of the state in delivering consumers' rights and choices'.

Retailers are involved in a number of policy areas such as preventive health policies that promote the consumption of fruit and vegetables where there is a clear convergence in the interests of retailers and government. However, food safety has been an area in which the shift of responsibility has been particularly extensive. The 1990 Food Safety Act was a significant point that made retailers give more consideration to their responsibility for food safety. 'The ability of the major retailers privately to regulate their supply chains and thus guarantee food standards has been formally recognised with the Food Safety Act (1990)' (Marsden *et al.*, 2000: 53). An important framework of food law still exists and may lead to enforcement action on retailers that sell food that is contaminated or unhealthy. Nevertheless, the general trend is for public food safety regulation to become 'less detailed and less prescriptive. At the same time, important forms of private regulation are more detailed, with a high degree of intervention curtailing freedom of regulated firms' (Havinga, 2006: 529). Thus, 'it is increasingly the quality and safety standards set by retailers and other companies, rather than those set by governments, which matter most to producers and consumers' (Food Ethics Council, 2005: 3). As a manager at Asda Wal-Mart has put it, 'The retailer has moved to being the guardian of food standards' (Brown, 2005: 20).

This form of relationship is entirely consistent with a depoliticized mode of governance where responsibility is shifted from government to other actors. 'In essence, depoliticisation as a governing strategy *is the process of placing at one remove the political character of decision-making*' (Burnham, 2001: 128). It has typically been discussed in relation to macroeconomic policy making, but has also been deployed in other spheres of policy. Thus, the operational slogan 'all power to the central bankers' has its echo in 'all power to the multiple retailers'. However, the blurring of public and private authority is not without its problems. In relation to pesticides in particular what has developed is a supplementary system of regulation by retailers that goes beyond the requirements of the state system. Pesticides are the most intensively regulated area of the food chain, but nevertheless remain an area of consumer concern that could frustrate policy objectives such as high levels of fruit and vegetable consumption as a preventive health measure. Research by the Institute of Grocery Distribution (IGD) found that 'The respondents saw the use of pesticides as an area of interest and concern and wanted to know what had been used on their food products' (IGD, 2002). The coexistence of public and private systems of regulation produces tensions between the system of private retail governance and more traditional components of the regulatory state.

The Drivers of Private Regulation

The involvement of supermarkets in creating private systems of regulation of pesticides is primarily driven by commercial considerations, although part of that is protection of reputation and the brand image. As Retailer D put it,

'We are working to protect our logo, the reputation we have in terms of food safety'. Supermarkets are relied on by their consumers as their most important provider of information on healthy eating, ranking higher than the government regulatory body, the FSA (FSA, 2007: 62). Pesticides residues on food remain a subject of concern for retailers. Hence, by demonstrating that their fruit and vegetables have fewer pesticide residues than those of their competitors, supermarkets may be able to develop a competitive edge. As one supermarket manager commented, 'If the customer wants no residues we should look on that as an opportunity … This fits in with the brand image, the message in our shops'. Another retailer commented: 'It is not just about consumers, it is about protecting the whole brand'.

Two notes of caution are necessary. Supermarkets are not a homogeneous group; they have their own perceptions of where they fit into the market and this in turn drives their strategies, particularly in terms of the extent to which they place an emphasis on price or quality considerations. It is important to note that research by the IGD 'shows that price, sell-by date and taste are the main factors influencing purchase by over 70% of consumers. In contrast less than a quarter of consumers consider factors covering production issues such as GM, animal welfare, and grown in the UK as influencing purchases' (IGD, 2002). Retailer A, one of the innovators in terms of private regulation, is rather unusual in stating that '[o]ur business is not about large volumes, it's about premium quality, doing things differently, a niche food retailer'. In contrast, Retailer G, one of the smaller firms in the sector, commented: 'One is always driven back to costs. The market place we've got is very competitive so you can't afford to be out of line commercially.'

The technical and managerial resources that supermarkets have to deal with these issues vary in their extent. Retailer C remarked of Retailer D, 'they haven't got the technical resources in house, they are looking for a stable risk management route'. Retailer D confirmed in interview that it was more reliant on the use of external consultants, commenting 'We need to compete in terms of how the rest of the industry is moving. We're a retailer, don't apply pesticides to anything, it's not our area of expertise.' Retailer F frankly admitted that it was replicating all the things that Retailers A and H had done. Retailer C noted that Retailer D 'are adopting the system we have, calling it a different name and slightly different, but I am more than happy with that'. Retailer G had experienced substantial staff cuts over the last year and commented: 'We let the regulators and our growers get on with it'. He elaborated: 'We are all different. My line manager wanted me to do a [Retailer A] list. I said I would have to sit behind my desk all day doing the list, updating it, that's why we have PSD.' Not all retailers can deploy the resources required for effective private regulation.

Second, retailers are not internally homogeneous. The lead on pesticide issues is usually taken by technical managers who have the relevant expertise. However, it was suggested in interviews that policy communication or marketing teams might have a different perspective. When Retailer F was asked if its new programme was intended to give it a commercial edge, the reply was: 'It doesn't give an edge at the moment. The head of commercial trading will dislike it ... I am not just interested in the price this year. I want a long-term sustainable relationship with packers and then with growers ... My bosses would say this is about giving us a commercial edge. I'm passionate about it.'

At Retailer C the role of technical managers for produce categories was 'to make sure we have products on the shelves that the customer wants in terms of quality and price all year round and feeding back into best farming practice'. In contrast, the trading law and technical manager acted as 'an overall coordinator and referee, making sure that the balance is right across board. [I'm] coming from a customer perspective and risk to human health and legal compliance.'

There may be limits to the extent to which it is possible to create a greener image for a supermarket. As one retailer commented, 'The ability to put oneself on high moral ground is shrinking, there is not much headroom left to say how squeaky clean we are, the likely public relations return is reducing, particularly on the produce side'. However, there are also more negative drivers about potential damage to brand and reputation. Concern about adverse media publicity was a recurrent theme in the interviews. As Retailer C put it, 'No one wants to be the person in the paper next time round'. Another retailer commented: 'Any retailer does not want a front page in tabloids ... protecting the brand a key part of that'. A senior technical manager commented: 'Directors won't say to you, I have a concern on X, it's only if it becomes an issue and it is up in lights in the local paper and [our] name is on television'. His counterpart in another retailer commented: 'All our residue testing is geared around public perception. If the *Daily Mail* rings up and says "You've got a problem with your spinach", press office can say, "We know about that and fixed it months ago" ... Pesticides management works fine if it doesn't get in the papers.'

NGOs and the negative publicity they can generate were also seen as drivers of company policy. One retailer had at one time refused to meet with them, but '[w]e had a lot of bad problems from campaign groups who started picking on [Retailer A]'. The retailer started a strategy of engagement with such groups, holding conferences to which they were invited along with government regulatory agencies. Retailer C noted that Retailer B, which sought to cultivate a green image, 'have lots of conversations with Friends of the Earth and Greenpeace'. Retailer H, which also seeks to portray a green image, noted that it had very good links into NGOs that sat on its group of experts on pesticides, specifically naming Friends of the Earth and the Royal Society for the Protection of Birds. One retailer considered that the pressure was lower than it had been. This was partly because some of the battles had been won in terms of retailers responding to concerns. It also reflected the increased importance of other issues including nutrition, waste and energy issues.

How Private Regulation Works

The exercise of private regulation in retailing largely takes place on a firm-by-firm basis with considerable variations in the schemes devised by each firm. This reflects the drive to gain a commercial edge over competitors and the 'company state' model of government–business relations in Britain, with trade associations often being relatively weak intermediaries and many initiatives taken by firms. The least well resourced retailer did, however, work closely with the Fresh Produce Consortium noting that it would have insufficient clout to work independently. Some use was made of informal networks to exchange information. One of the managers at Retailer D commented: 'We've all got contacts. It's an incestuous business. Individuals move around firms and develop contacts in that way. You've got your own contacts.' Retailer H commented: 'We've done bits and pieces with other retailers over the years. I think you know which ones we would work with.'

The role of standards organizations

Insofar as retailers do work together, it is through standards organizations that can set requirements about the way in which products are produced. In this way they can offer a collective response to consumer concerns and reassure consumers. In the UK the origins of these schemes go back to the NFU–retailer partnership of the early 1990s. EurepGAP (now GlobalGAP) was set up in 1997 by a German-based company EHI to enable retailers to cope with seasonality of fresh produce and the need for imports to provide the continuity of supply demanded by consumers. It has 500 members who are European food retailers, agricultural producers and associate members from the input and service side of agriculture:

> EurepGAP was driven by the desire to reassure consumers ... Food safety is a global issue and transcends international boundaries. Many EurepGAP players are global players in the retail industry and obtain food products from around the world. For these reasons a need has arisen for a commonly recognized and applied reference standard of Good Agricultural Practice which has at its centre a consumer focus.
>
> (EurepGAP, 2006)

Its role is to create international standards for agriculture. 'Technically speaking EurepGAP is a set of normative documents suitable to be accredited to internationally recognized standards such as ISO Guide 65' (EurepGAP, 2006). Its focus is on practices on the farm, with other Codes of Conduct and certification schemes relevant to food packing and processing coming into play once the product leaves the farm. It has met objections from African countries arguing that it was an additional trade barrier and the issue has been reviewed by the World Trade Organization.

EurepGAP covers imported produce, and as GlobalGAP increasingly trades around the world, while the UK Assured Produce Scheme was 'set up as a means of responding to requests from retailers on food safety' (interview with former Chair, 27 June 2005). Its importance is also recognized in the National Pesticides Strategy: 'The Farm Assurance Schemes have been and are likely to remain important vehicles for changing practice in the farming industry' (PSD, 2006: 12). If growers of produce want to market to the major multiple retailers, they have to be members. Fifty-two fruit, vegetable and salad crops are covered. The board and council of the scheme are made up of representatives of the UK supermarkets, growers, processors and the NFU. The scheme involves registering the crops grown, the completion of a Self-Assessment Questionnaire and an annual audit on the farm by an independent certifier to verify that the questionnaire has been properly completed. One of the objectives is to minimize pesticide inputs. It is claimed that the scheme 'saves growers substantial time and costs from having to meet many potentially different requirements from each of the market outlets' (Assured Food Standards, 2005). While this is the case in many respects, it does not entirely apply to pesticides, given the different requirements of different retailers.

Testing for residues

The primary safety/health concern in residues is exceeding the MRLs. All the retailers interviewed acknowledged that such occurrences are rare and are more frequent on imported produce. This contrasts with the presence of residues below the MRLs which through the PSD approval process are considered safe, but the consumer, perhaps stimulated by media commentary, considers as being unhealthy and/or unsafe. The broader context is 'a society characterised by a higher level of risk consciousness' producing what has been called a 'risk society', in which there is an 'increased recognition of the potentially negative effects of scientific and technical developments, the positive effects being increasingly taken for granted' (Lowe *et al.*, 2008: 228). These perceptions then feed back to the retailer in terms of what one retailer described as 'This is what our customers have told us ... they want'. This particular retailer then placed this information into its corporate responsibility framework, leading to the conclusion that what was required was 'food without the baggage of pesticide residues'. What was less clear was how the retailer would supply sufficient quantities of food at a competitive price.

Retailer schemes of pesticide classification

Retailers may prohibit the use of particular pesticides that have been approved under the state regulatory system or for others permit their use only with the specific consent of the retailer. There are a number of pesticides that are approved by CRD and prohibited on produce sold by a particular retailer. The basis of this list differs between retailers. Retailer H, which particularly emphasizes a green image, had prohibited approximately 130 and was monitoring another 300. As far as the monitored list was concerned, 'you can use them but you must have sound and solid reasons and information must be readily available. [It's] a way of tightening the screw, putting more pressure on people.' Retailer A started with a list of 79 pesticides it wanted to prohibit, but found that this would create difficulties with suppliers. However, it is putting continuous pressure on its suppliers: 'When we sit down with suppliers to review business as well as a commercial review we will have a technical overview, look at pesticides issues, what progress they've made'. Retailer F, who was following the lead of A and H, was compiling 'a restricted list to give suppliers a steer, step out of old fashioned chemistry'. A third-party organization was being hired to establish a matrix that could be used as a basis for decision making and was evaluating 260 pesticides in terms of that matrix.

However, Retailer C took a somewhat different stance, although it did have its own classification system:

> We won't go further than legislation in terms of we don't like X because Friends of the Earth don't. We don't like to work that way ... If you go too far, we'd have problems with availability, then people just switch and you've lost that trade, availability has to be good in [the] produce market.

Most retailers use a red (prohibited), yellow (requiring permission) and green (permitted) classification system for pesticides, but Retailer D thought that this was too coarse a classification and that a more elaborate risk management system would provide a good management tool for growers. Their categorization shows how elaborate a private regulatory scheme can be:

1. Red: very high risks, should not be used unless a full case can be made. Over a 5-year period, about 5% of pesticide actives used outside the UK would be classified as red.
2. Amber 1: would state risks associated with use, would agree action plan as to what suppliers should be doing to demonstrate management of risks, ask for residue data. About 10% of actives.
3. Amber 2: some risks, would keep situation under review. A smaller proportion of actives than in Amber 1.
4. Amber 3: some issues but nothing that one needs to be concerned about.
5. Green: no issues. Most actives would be either Amber 3 or Green.

How does all this look from the perspective of the grower? A large-scale Fenland grower is one of the biggest leaf salad suppliers in the country, specializing in producing lettuce, onions and chicories. They had 17 different customers and six or seven major different specifications; Retailer C had different specifications for two contracts. Although they sold their produce to intermediaries, they stated that '[a]t the end of the day we are really dealing with supermarkets'. Given that price differentials were converging, retailers were keen to get some other point of difference. We were shown two plots of lettuce being grown for different supermarkets; one had a deeper shade of red, which was thought to provide a competitive advantage. Retailer A was seen as driving a lot of innovation in food and as having a very sophisticated technical department. Retailer H was seen as trying to upgrade its produce 'in one fell swoop' and lacked a technical department. Retailer A was seen as 'a little more understanding. It depends on which representative you are talking to, different approaches in interpreting what is possible.' Nevertheless, their slogan was 'if you don't comply, you can't supply'. Supermarket customers make regular visits to the farm and Retailer A had deemed that the onion washing facility was no longer of a sufficiently high standard, illustrating how retailers can exert influence on investment decisions by growers.

All supermarket chains test a sample of the produce coming into their stores for pesticide residues. Even the least well resourced retailer tested for 328 active ingredients. Retailer C tested about 700 individual products a year. Testing was focused on products where it was thought there might be problems, e.g. lettuce grown under protection in December and January. In general, the greatest pesticide residue problems occurred with imports, particularly in soft fruits and exotics such as papaya or lychee. Retailer C commented: 'The real struggle for [the] retailer is exotics'. More familiar products that are imported can also give problems. Retailer C had a problem with potatoes from Egypt with several years of MRLs being exceeded. Because the soil in Egypt is sandy and thinner, a fungicide could get on to the tuber. An action plan was agreed with the grower, allowing reclassification of the pesticide from red to amber in the retailer's scheme. If the problem did not reappear after a few years, a green classification would be granted.

Given that there is a state system for monitoring and reporting on levels of pesticide residues, why do retailers undertake their extensive testing? One reason is simply to safeguard against liability claims. As one retailer commented, 'If we were taken to court because someone died or was seriously ill, we have to prove due diligence, that we have done everything in our power'. There is also concern, as regulatory systems becoming tighter, about growers, particularly outside the UK, using pesticides that are illegal. It is less easy to tackle these problems when they arise outside the UK, although retailers do make considerable efforts. For example, Retailer A visited suppliers in Latin America for business reviews that covered progress made on pesticide issues. In one case, as a result, three amber list pesticides were replaced by biopesticides.

Retailer C emphasized the importance of managing customer perceptions of the industry: 'If you don't tell the customer about pesticides they're quite happy. What the customer typically says is: "You're in charge of that, let me get on with, I don't want to know"'. The human health risk from pesticides was negligible 'but the public don't see it that way, it's all about managing the customer's trust'. Retailer F agreed that 'consumer expectations have to be better managed by us, trust has grown. This is very hard ground.'

Category management

In order to understand how retailers control the use of pesticides by suppliers, it is necessary to understand the basic principles of the system of category management. Retailer F commented, 'Interaction is with the packer rather than the grower. Category management is done for supermarkets by suppliers.

Category management is how supermarkets do things and I can't imagine we would go back on that.' Retailers have rationalized their supply base:

> The major supermarkets now deal with just a handful of suppliers in key product areas (potatoes, root vegetables, brassicas, salads, top fruit, stone fruit and soft fruit) and take every opportunity to pass responsibility (and associated costs) for quality control and procurement, storage and distribution upstream to their key suppliers, in return for which the chosen few are rewarded with volume growth.
>
> (Fearne and Hughes, 2000: 763)

In essence, responsibility is shifted down the food chain to the grower, while power, particularly buying power, remains with the supermarket. Asked if there were any benefits in being a lead supplier, a manager at Retailer C commented: 'Not really. There is workload on behalf of other suppliers [and us], they don't get any extra business. We are treating them as industry experts.' The need for the supermarket to retain control, particularly in terms of traceability, requires reducing the number of suppliers. As Retailer G explained, 'The only way to control is by restricting number of suppliers, gone from 200 to 60 in last three years. We have one supplier for all our potatoes, I know that I have full traceability.'

Retailers make considerable use of the technical expertise of suppliers, particularly in relation to pesticides as Retailer C explained:

> We have lead suppliers, [XYZ] for brassicas, because their technical team is very good, they understand pesticides far better than I do. I am a facilitator and coordinator ... On brassicas, we know strengths and weaknesses of our supply base on pesticides ... We know how strong they are on knowledge and application of pesticides.

The State and the Retailer Governance Systems of Regulation

As discussed in Chapter 6, the links between retailers and PSD were relatively weakly developed. Often they were confined to attendance at open meetings of the ACP or perhaps membership of the Pesticides Forum. Retailer D commented: 'We only interact with PSD if they want specific information from us'. Retailer C was quite critical of PSD, although his statement reflects the different perceptions and priorities of a commercial and regulatory organization:

> You could use the retailer muscle to help to move [things] forward. That's where PSD and DEFRA should be looking. No sign of movement by PSD, say we always do it that way, you've got to do it that way ... In all my dealings with PSD what really frightens me is that they have no real life experience on the farm or what the market is saying ... We give PSD a hard time about being more robust. We have a love–hate relationship with them at times.

Retailer A had adopted a more proactive approach to PSD and had had several meetings with them. This supermarket's senior technologist thought that PSD did their job well, although he thought that the FSA was a 'lot more customer focused'. However, in relation to the retailer's prohibitions list, PSD had argued 'They're not dangerous, why are you trying to do something different?' and Retailer A had replied 'We're addressing customer concerns'. Once again this shows the different focus and priorities of the regulatory agency and the supermarket.

Discussions with PSD confirmed that they had a dialogue with supermarkets but it was limited. Although regulators had to be guarded about what they said, it was evident that there was some unease about supermarkets banning pesticides that had been judged to be safe by the approvals process. Expressing a personal view, one regulator commented that 'it makes it difficult for us, giving the response that lots of pesticides are not safe'. Regulators considered that they had no influence over what supermarkets did: 'This is not something we can do anything about. All pesticides have gone through [the] approval process and shouldn't cause a problem. We can't say to Retailer A and Retailer H, you can't do this.' Another regulator commented: 'Retailers have their cake and eat it, they expect impossibly high standards from farmers who are forced into [the] situation of using them'. There is an evident tension between the role of a state regulatory agency whose task is to rule whether particular pesticides are safe for people and the environment and supermarkets driven by the concerns of their consumers which leads them to ban the use of products deemed safe by the regulator after a thorough process of scrutiny. As Retailer F commented when asked how much company policy was driven by the consumer, 'With pesticides it is less about what is safe, it is about fear factors'.

The Costs and Benefits of Private Regulation

These need to be considered from the perspective of the grower, the retailer, the consumer, the regulator and the environment. The shift of power within the food chain from growers to retailers has been reflected in keener prices combined with demands for higher quality, and controls on the use of pesticides by retailers are one aspect of that. This increases the complexity of the task faced by the grower and the demands placed on management time. This is made more difficult by a shortage of technically qualified individuals able to manage pesticide use, although this partly reflects the level of pay that the sector can afford. A Fenland grower that was visited had a technical staff of two, one of whom had been trained up from the yard when it was evident that she had potential. Many of the costs of the system of private regulation are thus borne by the grower. 'Therefore, when a retailer develops a strategy for sourcing more sustainable products, they as governors of the supply chain can push all compliance risks and costs down to the supplier' (Thankappan and Marsden, 2005: 56). This can have distributive consequences that marginalize smaller farmers as better capitalized farmers are more able to adapt to such requirements.

From the perspective of the retailer, imposing additional requirements on growers is a way of defending their reputation and brand, particularly against adverse publicity, which is a widely shared concern. For some retailers it is also a means of demonstrating that they are 'greener' than their competitors. As Retailer D put it, 'It's an excellent way of defending our image if pressure groups come to us and say, "What are you going to do about it"? Nothing comes to the shelves at the expense of the consumer, operator or the environment.' Retailer D claimed that it was more able to take decisions about pesticides than the UK Government because it had a better database.

The new role of the consumer

At a general level society has become more organized around consumption as an activity than production. There has been a set of broader changes in society towards 'reflexive consumption, whereby people think of themselves as active, discerning consumers whose choices contribute to their sense of identity'. Consumption choices, although structured by retailers, assume a new significance:

> The growth of affluence has led to a stress on personal development, and society too is re-oriented towards the values of individuality and self-expression. With the decline in the defining power of old economic and political forms – associated with workplace, class and nation – self-identity forming has shifted to spheres where individuals have direction and control. This leads to a growing personal focus on consumption and leisure activities and the cultural resources and goods that surround them.
> (Lowe *et al.*, 2008: 228)

Retailers see themselves as proxies for consumers, and FSA data show that consumers are concerned about pesticides and have a high level of trust in information provided by supermarkets. From a supermarket perspective, mainstream consumers have a 'limited understanding of pesticides' and display a 'lack of awareness of regulatory process'. More discerning consumers, mainly in the ABC1 categories, are seen as 'not trusting of regulatory authorities' (Brown, 2005: 18). 'It is clear that consumers have become slightly unnerved by the number and regularity of food scares which is in the main due to the increasing industrialization of food' (Brown, 2005: 20).

It might be argued that supermarkets have contributed to this process of industrialization. Nevertheless, as Retailer F put it, 'I'm very focused on the consumer; I'm not going to take any risks. I am paid by [F] to be focused on the consumer.' The retailer is, of course, focused on the consumer to ensure that it does not lose the consumer. The retailer is trying to limit damage to its customer base and not its individual customers. Any loss of reputation may drive customers to its competitors. It is this that is at the root of the retailer's concern. By developing a 'greener' image, a particular retailer hopes to gain a competitive edge over its rivals.

This portrayal of the retailer as the proxy defender of the consumer is, however, more contested than the claims of supermarkets might suggest. From one perspective, the role of consumer is an increasingly empowering one that gives the shopper more influence than he or she enjoys as a voter. '"Choice editing", whereby businesses make environmentally and socially sound sourcing decisions on behalf of their consumers, is therefore a particularly important trend' (Food Ethics Council, 2007: 4). From another perspective, the consumer, possessing limited information, is vulnerable to the marketing ploys of the retailer. There is a clear link here to the depoliticization narrative. One can extend Hay's point (2007: 128) that 'the marketization of electoral competition is in danger of reducing the electorate to a series of atomistic rational consumers who, as atomistic rational consumers will rationally disengage'.

Encouraging or at least permitting supermarkets to take on additional tasks in the regulation of food safety is consistent with Hay's account of what amounts to an abdication of responsibility by politicians and a transfer of difficult challenges to experts, informed by a public choice perspective that 'markets know best'. It might be argued that allowing supermarkets to construct their own supplementary system of 'regulation' is consistent with Lord Falconer's belief that 'depoliticization of key decision-making is a vital element in bringing power close to the people' (quoted in Hay, 2007: 93). It is also consistent with a view that sees citizens primarily as consumers of public services.

By 'voting' with their purchases, consumers are able to express preferences for products to be produced in a particular way or even boycott products from countries of whose policies they disapprove. Micheletti (2003) constructs an account of political consumerism in terms of a theory of individualized collective action that combines self-interest and the general good. Political consumerism 'represents actions by people who make choices among producers and products with the goal of changing objectionable institutional or market practices' (Micheletti, 2003: 2). It may even provide a basis for political mobilization: 'For example, in their search for good soap for their child they may meet other families in stores who have the same or other similar problems. The families may decide to pool their private worries and engage in very concrete, problem oriented, local networks' (Micheletti, 2003: 18–19). Given that the majority of food and grocery shopping is still undertaken by women, Micheletti sees 'political shopping' as a means of empowerment for women and relates this specifically to the case of pesticides. She sees women as generally more sensitive to risk and inclined to react more negatively to the use of pesticides on goods needed for their family. Micheletti is aware of the tensions between political consumerism and more conventional conceptions of regulation. She admits that it can be seen as part of a flight from politics and 'an ersatz for proper democratic politics and engagements, whose focus is the political system and government regulatory policy' (Micheletti, 2003: 160).

There is quite a big leap from convenient consumption to political engagement and empowerment, not least for women. As Birchfield comments (2005: 598), in Polanyi's vision '[m]arkets are embedded into society rather than society being submerged into markets'. Yet this is the fate that awaits the political shopper for whom the moment of liberation experienced when a packet of free-trade coffee is bought may be a transient illusion that changes neither consciousness nor power structures. As far as the specific case of pesticides is concerned, but also more generally, one of the difficulties with the notion of the political consumer is that of information asymmetries between the shopper and the rest of the food chain. As Thankappan and Marsden note (2005: 45), 'The danger of proliferating schemes by individual supermarkets, on top of the existing national assurance programmes and umbrella initiatives … is that the consumer will be left more confused than ever. Research undertaken by the National Consumer Council, shows that consumers are completely baffled by the range of schemes, logos and claims that surround the food industry.'

When consumers make their purchase decisions, they are choosing from a menu of goods provided by the supermarket. The decisions are often made under time pressure and are largely driven by price. Hence, short cuts to information on quality are of considerable value. For example, consumers influenced by policy considerations place considerable trust in produce certified as 'organic'. What they often do not realize is that a limited range of 'traditional' pesticides can be applied to organic produce grown in the UK: sulfur, soft soap and rotenone. These are either of natural origin (rotenone and soft soap) or simple chemical elements (sulfur) compared with the more complex substances typically used as pesticides in non-organic farming. Retailer F referred to the relative ignorance of consumers about organic produce: 'You get some daft responses. If you ask them about organics, they say no pesticides are required, when you explain there are pesticides applied, they get very upset.' Retailer H commented:

> In area of organics there is a perception that it is pesticides free … Do you realize that copper is used heavily and is more harmful than pesticides in current production? When this comes out sales of organics will collapse, [the] public will say 'we've been misled'.

The informed consumer using his or her purchasing power to affect policy, and perhaps even mobilizing with other consumers, is the exception rather than the rule. The typical consumer is an 'information taker' rather than a 'policy maker'. The choice of supermarket made by a consumer is influenced, apart from the fact that there are often a limited number of supermarkets to choose from within a given locality, by a mix of price and quality considerations encapsulated in the image cultivated by the supermarket. However, this is a marketing tool to entice the consumer within the store, not a source of empowerment.

From the regulator's perspective it is operating a rigorous system for the approval (and withdrawal) of pesticides. By prohibiting the use of approved pesticides, retailers are implying that the state system is less than entirely satisfactory, although retailers would claim that they are not trying to undermine it. As Retailer H put it, 'We are supportive of the regulatory system, what we have looked at is a slightly different approach … We are slightly ahead of the legislation, but it's a kind of positive approach as well.'

One way of assessing how the private regulatory efforts of retailers meet environmental needs is to look at how far they encourage the wider use of biopesticides by growers. In interviews retailers acknowledged the environmental benefits of biopesticides, but were reluctant to take a proactive role. A principal exception was Retailer A, which had discussed alternatives with growers and provided examples of three pesticides on its amber list replaced with bioproducts. It had also sought to develop links with companies producing biocontrol agents. However, in general, retailers did not seem to be very well informed about biological alternatives to synthetic pesticides, welcoming the visit from the project team as an opportunity to learn more. Indeed, in 2008 one retailer held a conference on the subject in which the project team gave a presentation as a means of disseminating the message about their potential contribution to growers. Retailers were understandably reluctant to recommend particular products to growers. As the environmentally oriented Retailer H commented, 'We have to be careful on chemicals and biologicals, make sure we don't start saying to the grower you should use product x rather than product y'. Retailer D commented: 'We always have to ask ourselves what effect does it have on us commercially in terms of costs and yields? We are always driven back to costs.'

Conclusions

Using their contracts with growers for leverage, large supermarket chains have constructed often elaborate systems designed to prohibit or restrict the use of pesticides permitted by the state regulatory system. 'The regulatory state model has been widely criticized for its centrist conception of state power' (Thankappan and Marsden, 2005: 56). The advantage of such a system is that its objectives are transparently stated and it seeks to apply regulations consistently to different actors in accordance with the stipulated goals. This is in contrast to the variability of standards between retailers that imposes additional costs on growers. There is also scope in the regulatory state for regulatory innovation in terms of facilitating the registration of more environmentally sustainable pest control agents such as biopesticides.

The private regulatory system developed by supermarkets can be portrayed as empowering the consumer. Supermarkets are attempting to respond to consumer concerns, but it is evident from our interviews that they are also trying to shape and manage those concerns. Above all, as profit-making organizations they are driven by a search for commercial advantage. A technical manager in a supermarket chain faces the challenge of doing 'all the work to ensure that the product is produced safely, protects the environment and biodiversity, maintains quality standards and availability at a price we need to sell it on to keep our profits'. If resources are devoted to promoting environmentally safer actives such as biopesticides, this is the result of a commercial positioning of a particular retailer like A. Initiatives are always going to be influenced by commercial considerations, not least because 'sustainability as a set of quality standards may provide leverage for large enterprises to control markets and raise barriers to competition' (Thankappan and Marsden, 2005: 56).

In some respects, the objectives of retailers are contradictory. They would like to see residue-free fresh produce, but this is inconsistent with the need for a continuous supply of competitively priced goods that meet a particular specification. Above all, there is a tension between this objective and availability of produce which is a major commercial driver as lack of availability can lead to loss of market share. Some of the most challenging residue problems arise with imported produce, but as Retailer C put it, 'We're still going to need strawberries at Christmas; it would be a brave retailer who said we're not going to do things out of season'.

A polemical literature has emerged criticizing the displacement of retailers in the contemporary British political economy (Blythman, 2005; Simms, 2007). However, although some specific retailer practices may be curbed by the Competition Commission and the proposed supermarket ombudsman, retail concentration and the leverage it gives them on the food chain are not going to go into reverse. 'The indication is that the market will concentrate further, with the large multiple stores increasing their shares through new openings, and possibly through mergers' (Clarke *et al.*, 2002: 154–155). What is needed is a more sophisticated debate about the roles of supermarkets that neither demonizes them nor portrays them as liberators of the consumer. In the specific case of pesticides regulation, there needs to be a more structured dialogue between the supermarkets and CRD.

Retailers are able to exert their substantial influence in the food chain to influence grower decision making in order to protect their brand reputation. It should be noted that there is considerable variation in the practices of retailers, influenced by the resources available to them, their analysis of their market position and their assessment of the social composition of their customers and their purchasing preferences. Some retailers are more proactive in relation to pesticides than others.

What in practice amounts to a supplementary approval system contributes to misconceptions over pesticide safety by in effect the retailers running their own approval system. This gives the perception that not all approved pesticides are safe. Their standards can be very difficult for growers to meet and can be contradictory. For example, they would encourage the use of natural predators but would reject any pre-packed foods that have insects in them.

The supplementary system of private governance for pesticides does pose some challenges. First, different retailers have different requirements, enhancing the complexity of decision making for growers and producing in effect what is a non-standardized system of pesticide use. Second, retailers prohibit rather than promote alternatives such as pesticides, although they would reasonably argue that they cannot endorse particular products. Third, retailers lack public accountability and their actions are often driven, quite legitimately, by a desire to gain an edge over their competitors rather than by considerations of public policy.

References

Assured Food Standards (2005) *Frequently Asked Questions*. Available at: http://www.assuredproduce.co.uk/AProduce/faq.asp (accessed June 2005).

Birchfield, V. (2005) José Bové and the globalisation countermovement in France and beyond: a Polanyian interpretation. *Review of International Studies* 31, 581–598.

Blythman, J. (2005) *Shopped: the Shocking Power of British Supermarkets*. Harper Perennial, London.

Brown, C.M. (2005) Retail perspectives on crop production. In: *The BCPC International Congress Proceedings: 2005*, Vol. 1. British Crop Protection Council, Alton, UK, pp. 17–22.

Burnham, P. (2001) New Labour and the politics of depoliticisation. *British Journal of Politics and International Relations* 3, 127–149.

Clarke, R., Davies, S., Dobson, P. and Waterson, M. (2002) *Buyer Power and Competition in European Food Retailing*. Edward Elgar, Cheltenham, UK.

EurepGAP (2006) *Eurepgap Protocol for Fresh Fruit and Vegetables*. Available at: http://www.eurepgap.org/Languages/English/about.html (accessed June 2006).

Fearne, A. and Hughes, D. (2000) Success factors in the fresh produce supply chain: insights from the UK. *British Food Journal* 102, 760–772.

Food Ethics Council (2005) *Power in the Food System: Understanding Trends and Improving Accountability*. Food Ethics Council, Brighton, UK.

Food Ethics Council (2007) *'Food Miles' or 'Food Minutes': Is Sustainability All in the Timing?* Food Ethics Council, Brighton, UK.

FSA (2007) *Consumer Attitudes to Food Standards*. Food Standards Agency, London.

Havinga, T. (2006) Private regulation of food safety by supermarkets. *Law and Policy* 28, 515–533.

Hay, C. (2007) *Why We Hate Politics*. Polity, Cambridge, UK.

IGD (2002) *Price is key concern for UK consumers*. Available at: http://www.igd.com/cir.asp?cirid=370&search=1 (accessed June 2006).

Lowe, P., Phillipson, J. and Lee, R.P. (2008) Socio-technical innovation for sustainable food chains: roles for social science. *Food Science and Technology* 19, 226–233.

Marsden, T., Flynn, A. and Harrison, M. (2000) *Consuming Interests: the Social Provision of Foods*. UCL Press, London.

Marsden, T., Lee, R., Flynn, A. and Thankappan, S. (2009) *The New Regulation and Governance of Food: Beyond the Food Crisis?* Routledge, Abingdon, UK.

Micheletti, M. (2003) *Political Virtue and Shopping: Individuals, Consumerism and Collective Action*. Palgrave-Macmillan, Basingstoke, UK.

PSD (2006) *Pesticides and the Environment: a Strategy for the Sustainable Use of Plant Protection Products*. Pesticides Safety Directorate, York, UK.

Simms, A. (2007) *Tescopoly*. Constable, London.

Thankappan, S. and Marsden, T. (2005) *The Contested Regulation and the Fresh Fruit and Vegetable Sector in Europe. Working Paper Series No. 27*. Centre for Business Relationships, Accountability, Sustainability and Society, Cardiff University, Cardiff, UK.

8 Conclusions

This book originated from a UK Research Council-funded RELU programme on the Rural Economy and Land Use (RELU) which attempted to address the frustrations of natural scientists who struggle with the apparent difficulty of bringing biopesticides to market when their biological intuition suggests they are such obvious solutions to pest problems. In an attempt to unravel the problem and come forward with solutions, natural and social scientists joined forces. The social scientists fell into two groups. Economists focused on a series of economic barriers to adoption of biopesticides, while political scientists considered the system for their regulation is principally at fault. As we have seen in the preceding chapters it is clearly more complex than that. In this concluding chapter we try to draw together some of the emerging thoughts.

There is a very large number of plant species cultivated throughout the world that are used to feed the human population and their livestock, for fibres and building materials or for fuel. We have focused primarily on food crops but the emerging principles apply to all crops. A relatively small number of plant species such as rice, wheat and maize are cultivated to produce the staple foods for the world population, while a huge number of plant species provide the greater nutritional variety and interest in diets. The yield or quality of each of these plant species, whether or not it is a staple, is limited at some stage during its cultivation by a community of invertebrate pests, diseases and weeds, collectively referred to as pests. The increasing world population and its increasing affluence and urbanization are placing ever greater strain on the world food production system to produce and distribute more and more food in a sustainable way to nourish that population adequately. Increased food production is achieved first through improvements in agronomy to achieve greater yields per unit area and second by minimizing waste and spoilage. The control of pests in the field contributes to the maximization of yield and reduces spoilage postharvest. The current

© CAB International 2010. *Biopesticides: Pest Management and Regulation*
(A. Bailey et al.)

debate over the risk to food security is pushing all these issues higher up the political agenda. Regulation needs to take account of the benefits as well as the costs of growing crops using plant protection products.

The solution to an impending food security crisis is potentially at odds with the policies on sustainability that propose that mankind treads lightly on the environment, does not aggravate and ideally mitigates climate change and its impacts, and does not deplete biodiversity further. Modern large-scale agriculture does have the potential – with large machinery, modern plant varieties and agrochemicals – to feed the world, but at what cost? It has been set out earlier how crops are complex ecological communities and that the manipulation of different components of these communities to the benefit of the crop and of mankind should be approached as an exercise in applied ecology, rather than the application of technology despite the ecology.

Each crop pest also has its own community of predators, parasites and diseases that limit their numbers and vigour. Historically mankind has relied largely on these interactions to achieve some control, yet devastating crop losses occurred from time to time. Over the last 100 years there have been major advances in our understanding of the complex interactions between the crops we cultivate and their biotic environment and of the potential to manipulate this environment to the benefit of the crop. At the same time a sophisticated chemical industry has developed that discovers, manufactures, distributes and sells pesticides that control many crop pests. Coincident with this some regions of the world, particularly the developed North, have experienced dramatic globalization of food markets, initially of the staple crops and recently of nearly all crops. This has politicized trade and regulated aspects of food production. At this point further challenges emerge due to the different spatial and temporal scales of the political and economic landscape.

The rapid growth of the global agrochemical industry after World War II focused effort on products for crops grown on large acreages (mainly staple and fibre crops) where sales would be large and profits maximized. This remains the case to this day, but with the hope or expectation these products will also find use on at least some of the very many 'minor crops', many of which are significant contributors to a balanced diet. In the developed world the sale and use of pesticides are more closely regulated than for any other chemicals including pharmaceuticals. The stringent conditions to which pesticides have to comply have evolved as our understanding of the environmental and human health impacts of these chemicals in particular have developed. Initially large numbers of products emerged which found their way into every corner of agricultural production. As regulations have matured and become tighter, the pipeline of new chemistry has declined at the same time as products are being withdrawn from the market as they no longer meet the more stringent regulations or are now uneconomic to produce and market. The consequence is there remain chemical solutions, though a more limited choice, to many pests on the broad-acre staple crops, but many minor crops now have very few chemicals available to control their pests, especially given that a product may not work well on some soil types.

Though these differences have always been apparent since pesticides were first invented, they are now more evident than ever.

Intuitively, the manipulation of all the biotic interactions within the cropping environment, along with the judicious use of pesticides where they are available, in IPM must be a solution. This is what the revised European regulatory regime enacted in 2009 is based upon. Yet this is where the challenges really are, and there are many of them manifesting themselves on different scales. Many biological solutions require detailed research to develop the 'biological fix', but also considerable on-farm management skill to implement. Second, biological interactions are specific to two or at most a very small number of species, so novel solutions need to be developed for almost every crop–pest combination. This contrasts greatly with the chemical approach where 'modes of action' dictate that a chemical will be effective against a far wider range of pests. The biological solutions are very resource intensive while only delivering niche solutions, which the economics of the global agro-chemical industry finds it difficult to reconcile. Additionally the regulatory process has evolved to accommodate the chemical model and is less easily applied to a multitude of niche biological solutions.

The IPM approach will include a number of pest management methods in a portfolio that in many cases may be complements, or substitutes. Here, complementarity between techniques may take the form of relationships that are simple additive (site and time specific), spatial additive (field margin to centre, ground level to crop canopy activity, etc.) or temporal additive (early to late season active), and any of these relationships may serve to increase mean levels of pest control efficacy. Furthermore, the combination of technology sets that include functional substitute techniques,[1] by building resilience into systems, could prove highly effective at controlling the variance of overall pest control function.

The acceptance of the portfolio concept in which a number of control options can be integrated together in different combinations to deliver an equivalent outcome has to be complemented by a high level of knowledge and management skill to implement in an economically acceptable way. Some technologies group together due to their similarity of approach and methods of application. A farmer may therefore adopt a number of approaches that are similar, but be constrained by the economics of application and a risk-averse nature from stepping into the next technology portfolio. This alone may constrain the step away from a chemistry-based approach to pest control into an IPM strategy with many components.

Individual farmers may struggle to acquire and interpret the knowledge required to implement complex portfolio approaches. In much of the developed world there is declining investment in agricultural education and research, and frequently the complexities of the technologies involved are becoming greater than the individual small business can absorb. In developed countries there is increasing polarization of farm business size with small family farms struggling to accommodate technological advances and deliver a profit. This contrasts with those farm businesses that are expanding through amalgamation and takeover and relish new technology, particularly in the IPM arena. Their large size enables them to employ the specialist technical personnel who understand and implement the developing and increasingly complex technology.

We have therefore arrived at a failure in both the regulatory process and traditional economic models that cannot accommodate satisfactorily biological products, including biopesticides, for application to minor crops. So perhaps biopesticides are 'niche solutions to niche problems'. This is compounded further by the nature of the market for minor crops, many of which are the fruit and vegetables of our diet that are sold fresh with exacting quality standards. Individual units of harvest of predefined size, such as an individual apple or a head of lettuce, are marketed. The production of staple crops from broad-acre agriculture is frequently processed prior to human consumption or is used as animal feed or non-food crops, so total yield is of greater importance with the quality of individual units of crop being subordinate to it. The market price is determined by the global commodity markets.

To address the minor crops market, a collection of new biological technology companies has emerged since the early 1990s in an attempt to deliver to these niche markets, frequently in clearly defined and modest geographical regions. These companies are characterized as SMEs, although in practice they are often micro enterprises with just a few employees and are frequently driven by biological entrepreneurs with a passion for what they believe biological solutions can deliver. These companies invariably have very limited budgets and individually very little influence by comparison with the more global chemical industry. Some of these small companies have formed the IBMA to give themselves more political influence and a role as stakeholders in policy debates. However, the market size for individual products will invariably remain small, with a few exceptions, due to the minor crops at which many products are targeted.

The targets in Europe for many of the biological products are pests of 'minor' fruit and vegetable crops which have largely been grown in an unsubsidized agricultural economy, subsidies having been directed primarily towards broad-acre staple crops and livestock. Consequently producers of minor crops have developed a much greater awareness of and a close working relationship with their markets, particularly the multiple retailers. This reveals yet another David and Goliath relationship. The sale of fresh produce by multiple retailers is very profitable. We have seen through our interviews with retailers their sensitivity (driven in part by the financial sector) to maintenance of market share, defence of brands and reputation as well as shareholder confidence, all of which could be adversely affected through marketing of inferior-quality fresh produce. Of particular importance to them is the detection of pesticide residues on produce or the presence of any pest or 'foreign bodies' with the produce. Retailers apply considerable pressure on their suppliers to ensure pesticide residues are not detected. Some individual producers and marketing groups may make considerable investments in technology and skilled management resources to manage pesticide applications with great care and to incorporate biological solutions to pest control into their production process, but the risk remains theirs and not the retailers'. The retailers, with a few exceptions, do not concern themselves with the regulation and promotion of pesticides and biological alternatives, leaving it to the manufacturers and the farmers to drive the need for products for use on minor crops. However, the pesticide usage constraints placed on suppliers by some retailers, particularly the need to select chemical control options from a restricted list of pesticides, does create a private system of regulation which supplements and perhaps undermines the public system.

The regulators in Europe and the USA do seem to be rising to the challenge to facilitate the registration of biopesticides by establishing specialist teams to share their knowledge with manufacturers to ease the collation of all relevant data. But could biopesticides be a perfect substitute for chemical pesticides? Arguably, no! Their efficacy, as measured by percentage kill, their speed of action, as measured by time to achieve 50% kill, and their reliability, as measured by the variance of their kill rate, are generally far less attractive than the chemical alternative. Sole use of biopesticides in a mono-technology approach would probably be economically suboptimal at least in private cost terms. More generally, the UK ACP (2003) report states that:

> Many of the alternative control tactics are only partially effective or have a very selective action against certain pests. They may therefore give inadequate control when used alone.

While used in isolation, the efficacy of many IPM components does appear less attractive than that of chemical control. One may therefore caution against the search for biopesticides as 'silver bullets' and suggest that combined, or integrated, systems approaches are more suitable. There is evidence that efficacy of IPM improves with an increased number of biocontrol options used together.

Biopesticides could prove to fulfil a number of key roles in IPM portfolio approaches. However, consideration of portfolio technology adoption, particularly in the case of mixes of knowledge or system technologies (sometimes referred to as disembodied technology) and capital or material product technologies (referred to as embodied technologies), is complex. This distinction is important for two phases of the technology development process. First, only embodied technologies are likely to be brought forward to market by private companies. While private companies can register patents and capture rents from the sale of products which embody a technological advance, individual users are not excludable from disembodied knowledge technologies. Second, private companies have much to gain from the extension and market development of embodied technologies while only government and NGOs are likely to promote system or disembodied technological advances but both probably lack the resources for the task. Biopesticides, alongside pheromones, plant extracts and antifeedants, can be considered as embodied technologies and might find their commercial champions. However, other disembodied IPM portfolio technologies are unlikely to be championed by the private sector unless those companies recognize the true portfolio nature of IPM upon which their own sales rely. The adoption problem becomes more severe when there are both scale economies and network externalities in the technology adoption complex.

There has been a considerable decline in the number of active ingredients available to the broad-acre farmer although there are generally still many chemical solutions available. Many minor crops are in a very different position with no chemical solutions for many pests. Through economic necessity, growers have been driven to use biological alternatives. A good example is the tomato and cucumber growers of north-west Europe, who grow their crops in glasshouses. Initially resistance to acaricides drove the introduction of predatory mites to control the two-spotted spider mite. There was then a need for further biological solutions for other pests, but the real economic push for complete biologically based integrated control came with the introduction of bumblebees to pollinate the crop, which gave even fruit set. Biological solutions were then driven by the need to preserve the valuable pollinators. This illustrates the need for the appropriate economic drivers at the scale of the farm if biological solutions are to be deployed.

A further scale consideration is the availability of technologies to produce adequate quantities of biopesticdes, or other biological control products, for the application to millions of hectares of broad-acre crops. This is well illustrated by the control of the soybean caterpillar with *Anticarsia gemmatalis* NPV in Brazil, which proved to be highly effective, but the inability to produce sufficient quantities of virus *in vivo* has put the whole technology and its wider reputation at risk. Therefore, until there are major technological advances that can shift the production of more biological products from *in vivo*-based systems to those of industrial chemical production, their use may be 'trapped' within minor crops.

The prime focus of the discussion so far has been to make suitable products, including biological agents, available for use, particularly on minor crops. The role this has in the wider delivery of a secure food supply is less clear. Government and regulator objectives have centred on ensuring products are safe to human health and cause no damage to the environment while to date taking less account of delivering food security or safe water supplies, although the agenda is rapidly shifting. It opens up the question of what agricultural land is for. Does it have functions beyond the supply of food? The answer to this is certainly yes. Land cultivation impacts on the release and sequestration of greenhouse gases, modifies the flow of water through the water cycle, alters biodiversity to name just a few. To deliver all these requires a greater joining up of the regulatory processes so land managers can deliver all we expect from land.

Compared with the macro agendas of climate change and food security, the micro agenda of biocontrol products may seem to be relatively insignificant. Nevertheless, they offer a key element of the IPM approach that is now at the centre of the EU regulatory agenda and is likely to play an important role in climate change adaptation, a topic that has been relatively neglected at EU level compared with climate change mitigation. The future development of the Common Agricultural Policy is also likely to prove an important context for the development of sustainable agriculture, for example in relation to the signals provided by agri-environmental schemes that provide an economic stimulus for the introduction of new technologies.

This book has also provided a lens to examine some wider issues in contemporary society. Can regulators overcome their inherent risk averseness to facilitate the development of new products which offer more sustainable solutions? Our research suggests that they can, given a suitable set of conditions and stimuli. In a world in which power has flowed down the food chain to retailers as proxy representatives of the consumer, how can they be encouraged to use their market power to support sustainable solutions? Part of the response to this challenge would involve their more effective integration in the relevant policy networks. Above all, our research has shown that collaboration between economists, political scientists and biological scientists can generate insights and understandings that would not be provided by work based on just one discipline.

Note

[1] Functional substitutes are often, rather derogatorily, referred to as functional redundance in the applied ecological literature.

Reference

ACP (2003) *Final Report of the Sub-group of the Advisory Committee on Pesticides on Alternatives to Conventional Pest Control Techniques in the UK: a Scoping Study of the Potential for Their Wider Use*. Advisory Committee on Pesticides, York, UK.

Index

Acarina 23–25
Adelges tsugae 47
Advisory Committee on Pesticides (ACP) 4
Africa
    integrated pest management 60–62
    push–pull strategy 108
African palm weevil 108
agri-environmental policy (AEP) 135–144
alien species 45–48, 79–80
    Diabrotica virgifera 46–47
allelochemicals 101–102
Anticarsia gemmatalis 87, 221
aphids 21–22
    cotton aphid 89
arable crops 59–60
Arthrobotrys species 95
Ascomycota 88–90
Aspergillus flavus 28
augmentation biological control 77–79
    inoculation approach 78, 79
    inundation approach 77–78

Bacillus 94
    subtilis 94, 95
    thuringiensis (Bt) 58, 78, 82–83, 104–106
        genetically modified crops 83–86
        safety 117
bacteria 29, 82–83
Beauveria bassiana 89, 104, 105–106, 117–118, 120
Bemisia tabaci 22
bioherbicides 97–98
biological control agents 52, 62–64, 73–75
    microbial agents 75–76
        invertebrate control 81–91
        nematode control 95
        plant pathogen control 92–97
        regulation 108–116
        safety 116–120
        strategies 76–81
        weed control 97–98
    natural compounds 51, 99
        microbial 100–101
        plant derived 99–100
        safety 120–121
        semiochemicals 103
    natural enemies 37, 72, 73–75, 104
        escape hypothesis 47–48
        harlequin ladybird 119
        strategies 76–81
    strategies 76–77
        augmentation biological control 77–79
        classical biological control 79–80

biological control agents *continued*
   strategies *continued*
      conservation biological control 80–81
      introduction 79–80
biopesticides 2, 3–5, 52, 71–73
   adoption 4, 198–199
   integrated pest management 103–108, 219–221
   microbial agents 75–76
      invertebrate control 81–91
      nematode control 95
      plant pathogen control 92–97
      regulation 108–116
      safety 116–120
      strategies 76–81
      weed control 97–98
   natural compounds 51, 99
      microbial 100–101
      plant derived 99–100
      safety 120–121
      semiochemicals 103
   policy networks 177–178, 194–195
      biopesticides 178–179
      change 182–187
      components 179–182
      institutional reform 193–194
      regulatory innovation 187–192
   regulation 5–8, 72–73, 108–116, 148–149
      biopesticides 172–174
      CRD 6, 8, 9, 135, 149–153, 154–158, 161, 173–174, 179–194, 208–209
      EU 5, 80, 158–169, 182
      innovation 8–10, 187–192
      OECD 171–172
      policy networks 177–195
      retail governance 198–199
      UK 6, 8, 149–158
      USA 8, 169–171
      WHO 171–172
   retail governance 201–204, 210–214
      cost 209–210
      pesticide classification 205–207
      residues 205
      standards organizations 204–205
      supply chain 207–208
   safety 114–118
      botanicals 120–121
      microbial agents 116–120
      semiochemicals 120–121
biostimulants 155–156
Black's five worlds 187–191
botanicals 99–100
   safety 120–121
Brazil 87
broad-acre crops 59–60
broad-spectrum activity 36–37
Bt crops 83–86

Canada 87–88
carbamates 34
cassava 90–91
Chemicals Regulation Directorate (CRD) 6, 8, 9, 135, 149–151, 161, 173–174
   biopesticides scheme 157–158
   cost 158
   efficacy testing 156–157
   enforcement 154–156
   institutional reform 193–194
   maximum residue limits (MRLs) 152–153
   minor use crops 151–152
   off-label approvals 152
   policy networks 179–182
   regulatory innovation 187–191
      pressures 191–192
   retailers 208–209
China 86
Claviceps 27–28
climatic zones 161–162, 164–166
Coleoptera 21
Collego 97–98
Colletotrichum gloeosporioides 97–98
Colorado potato beetle 23, 105–106
competition 43–44
Coniothyrium minitans 105
control 1–3
   biopesticides 2, 3–5, 52, 71–73
      adoption 4, 198–199
      integrated pest management 103–108, 219–221
      microbial agents 75–98
      natural compounds 51, 99–101, 103
   integrated pest management (IPM) 3, 16, 39–41, 62–64, 144–145, 218–221
      adoption 53–55, 131–135

agri-environmental policy
  135–144
biological control agents 52,
  62–64, 73–75
biopesticides 75–76, 103–108,
  219–221
broad-acre crops 59–60
crop protection 49–50, 53
developing countries 60–62
economic thresholds 48–49
field vegetable crops 57–58
genetically modified crops
  59–60
greenhouse crops 55–57
microbial agents 75–98
natural compounds 51, 99–101,
  103
natural enemies 37, 38, 47–48,
  72, 73–81, 104, 119
orchard crops 58
population dynamics 41–49
pesticides 32–33, 216–218
  carbamates 34
  development 33–35, 39
  dichlorodiphenyltrichloroethane
    23, 34
  fungicides 35
  herbicides 35
  organochlorine 34
  organophosphate 34–35
  problems with 35–38
  pyrethroids 34, 100
  withdrawal of 39, 160
cotton
  aphids 89
  genetically modified (GM) crops
    59–60, 84, 86
  integrated pest management (IPM)
    61
cottony cushion scale 79
crop protection 49–50, 53
  biopesticides 2, 3–5, 52, 71–73
    adoption 4, 198–199
    integrated pest management
      103–108, 219–221
    microbial agents 75–98
    natural compounds 51, 99–101,
      103
  cultural practices 50–51
  genetic methods 52–53
  greenhouse crops 55–57

integrated pest management (IPM)
  3, 16, 39–41, 62–64, 144–145,
  218–221
  adoption 53–55, 131–135
  agri-environmental policy
    135–144
  biological control agents 52,
    62–64, 73–75
  biopesticides 75–76, 103–108,
    219–221
  broad-acre crops 59–60
  crop protection 49–50, 53
  developing countries 60–62
  economic thresholds 48–49
  field vegetable crops 57–58
  genetically modified crops
    59–60
  greenhouse crops 55–57
  microbial agents 75–98
  natural compounds 51, 99–101,
    103
  natural enemies 37, 38, 47–48,
    72, 73–81, 104, 119
  orchard crops 58
  population dynamics 41–49
  pesticides 32–33, 216–218
    carbamates 34
    development 33–35, 39
    dichlorodiphenyltrichloroethane
      23, 34
    fungicides 35
    herbicides 35
    organochlorine 34
    organophosphate 34–35
    problems with 35–38
    pyrethroids 34, 100
    withdrawal of 39, 160
  physical methods 51
  plant breeding 51–52
  selective pesticides 50
Crop Protection Management Plans
  (CPMPs) 138–139
Cuba 106–107, 114
cultural practices 50–51
customers 210–213
Cydia pomonella granulovirus (CpGV)
  87, 105

Denmark 167–168
developing countries 60–62

DeVine 97–98
Diabrotica virgifera 46–47
dichlorodiphenyltrichloroethane (DDT) 23, 34
Diptera 21
disease 25–26
    bacteria 29
    fungi 26–28
    microbial control 92–93
        antagonism 93–95
        induced host resistance 95–97
    nematodes 30
    oomycetes 28–29
    Phytophthora infestans 30–31, 48
    vectors 21–23, 24
    viruses 29, 86–88
        vectors 24
downy mildews 28–29

economic
    integrated pest management (IPM) adoption 131–135
    thresholds 48–49
efficacy testing 156–157
entomopathogenic nematodes 74–75, 91–92
Entomophaga maimaiga 90
Entomophthorales 88–89
Entry Level Stewardship scheme (ELS) 137–144
environmental conditions
    pest population dynamics 43, 45
environmental persistence 36
Environmental Stewardship scheme 137–144
ergot 27–28
essential oils 99–100
EU
    genetically modified (GM) crops 83–85
    regulation 5, 80, 158–159
        biopesticides 166
        climatic zones 161–162, 164–166
        Denmark 167–168
        directive 91/414 159–160, 162–163
        EFSA 159–160
        EPPO 161
        loss of active ingredients 160
        Netherlands 168–169
        policy networks 182
        reform 162–169
        shortcomings 160–162
        Sustainable Uses Directive 163–164
        Water Framework Directive 167
European Crop Protection Association (ECPA) 154, 163
European Food Safety Authority (EFSA) 159–160
European and Mediterranean Plant Protection Organization (EPPO) 161
European spruce sawfly 87–88
experimentation 11–15
    model systems 13–15

field vegetable crops 57–58
Food Safety Act (1990) 201
forecasting
    environmental conditions 43
Frankliniella occidentalis 22, 38
fungi 88
    control 35
    entomopathogenic 88–91, 117
    pathogens 26–28
fungicides 35

genetically modified (GM) crops 59–60
    Bt 83–86
Gilpinia hercyniae 87–88
GlobalGAP 204–205
glyphosate 59–60
government policy
    agri-environmental policy (AEP) 135–144
    innovation 8–10
    policy networks 177–178, 194–195
        biopesticides 178–179
        change 182–187
        components 179–182
        institutional reform 193–194
        regulatory innovation 187–192
    regulation 5–8, 72–73, 108–116, 148–149
        biopesticides 172–174

Index

CRD 6, 8, 9, 135, 149–153, 154–158, 161, 173–174, 179–194, 208–209
    EU 5, 80, 158–169, 182
    innovation 8–10, 187–192
    OECD 171–172
    policy networks 177–195
    retail governance 198–199
    UK 6, 8, 149–158
    USA 8, 169–171
    WHO 171–172
  retailer power 200–201
greenhouse crops 55–57
gypsy moth 90

Hampton review 194
harlequin ladybird 119
Hemiptera 21, 22
hemlock woolly adelgid 47
herbicides 35
  genetically modified (GM) crops 59–60
Heterorhabditis 74, 105
Higher Level Stewardship scheme (HLS) 137
human safety 32, 38, 115–116
Hymenoptera 21

Icerya purchasi 79
India 61
induced host resistance 95–97
industrial farming 33–35, 216–218
insects 20–23
  aphids 21–22
  Colorado potato beetle 23
  cotton aphids 89
  cottony cushion scale 79
  European spruce sawfly 87–88
  insecticide resistance 37–38
  locusts 21, 22, 89–90
  microbial agents 81–82
    bacteria 82–83
    Bt genetically modified crops 83–86
    fungi 88–91
    viruses 86–88
  natural compounds 51, 99
    microbial 100–101
    plant derived 99–100
    semiochemicals 103
  natural enemies 37, 72, 73–75, 104
    escape hypothesis 47–48
    harlequin ladybird 119
    strategies 76–81
  pesticides 32–33, 216–218
    carbamates 34
    development 33–35, 39
    dichlorodiphenyltrichloroethane 23, 34
    organochlorine 34
    organophosphate 34–35
    problems with 35–38
    pyrethroids 34, 100
    withdrawal of 39, 160
  sawflies 21
  spittlebugs 89
  thrips 21, 22
  whitefly 22
integrated pest management (IPM) 3, 16, 39–41, 62–64, 144–145, 218–221
  adoption 53–55, 131–135
    agri-environmental policy 135–144
  biological control agents 52, 62–64, 73–75
  biopesticides 75–76, 103–108, 219–221
  broad-acre crops 59–60
  crop protection 49–50, 53
  developing countries 60–62
  economic thresholds 48–49
  field vegetable crops 57–58
  genetically modified (GM) crops 59–60
  greenhouse crops 55–57
  microbial agents 75–76
    invertebrate control 81–91
    nematode control 95
    plant pathogen control 92–97
    strategies 76–81
    weed control 97–98
  natural compounds 51, 99
    microbial 100–101
    plant derived 99–100
    semiochemicals 103
  natural enemies 37, 38, 72, 73–75, 104
    escape hypothesis 47–48
    harlequin ladybird 119
    strategies 76–81

integrated pest management *continued*
   orchard crops 58
   population dynamics 41–42
      competition 43–44
      economic thresholds 48–49
      environmental conditions 43, 45
      life history strategies 42–43
      predation 44
      size increase 44–48
intensive farming 33–35, 216–218
interdisciplinary working 10–11
   complex systems 12–13
   experimentation 11–12
   individualistic fallacy 13–15
   Oedipus effect 11
International Biocontrol Manufacturers' Association (IBMA) 155–156, 162–163, 164, 179–181
invertebrates
   insects 20–23
      aphids 21–22
      Colorado potato beetle 23
      cotton aphids 89
      cottony cushion scale 79
      European spruce sawfly 87–88
      insecticide resistance 37–38
      locusts 21, 22, 89–90
      pesticides 32–35, 39, 100, 216–218
      sawflies 21
      spittlebugs 89
      thrips 21, 22
      whitefly 22
   microbial agents 81–82
      Bt genetically modified crops 83–86
      bacteria 82–83
      fungi 88–91
      viruses 86–88
   mites 23–25
   molluscs 25
   natural compounds 51, 99
      microbial 100–101
      plant derived 99–100
      semiochemicals 103
   natural enemies 37, 72, 73–75, 104
      escape hypothesis 47–48
      harlequin ladybird 119
      strategies 76–81
Irish famine 30–31

Lepidoptera 21
Leptinotarsa decemlineata 23, 46
locusts 21, 22, 89–90
Lymantria dispar 47, 90

Magnaporthe grisea 96–97
Mahanarva species 89
maize
   genetically modified (GM) crops 59, 83–86
   western corn rootworm (WCR) 46–47, 84
maximum residue limits (MRLs) 152–153
Metarhizium anisopliae 89–90, 117–118
microbial agents 75–76
   invertebrate control 81–82
      Bt genetically modified crops 83–86
      bacteria 82–83
      fungi 88–91
      viruses 86–88
   nematode control 95
   plant pathogen control 92–93
      antagonism 93–95
      induced host resistance 95–97
   regulation 108–116
   safety 116–120
   strategies 76–77
      augmentation biological control 77–79
      classical biological control 79–80
      conservation biological control 80–81
      introduction 79–80
   weed control 97–98
mites 23–25
model
   integrated pest management (IPM) adoption 133–135
   systems 13–15
molluscs 25

natural compounds 51, 99
   microbial 100–101
   plant derived 99–100
   safety 120–121
   semiochemicals 101
      allelochemicals 101–102

Index

pheromones 102–103
natural enemies 37, 72, 73–75, 104
    escape hypothesis 47–48
    harlequin ladybird 119
    strategies 76–77
        augmentation biological control 77–79
        classical biological control 79–80
        conservation biological control 80–81
        introduction 79–80
neem oil 99–100
nematodes 30
    entomopathogenic 74–75, 91–92
Neozygites species 81, 89, 90–91
Netherlands 168–169
nicotine 99
northern jointvetch 97–98

Oedipus effect 11
oomycetes 28–29
orchard crops 58
Organic Entry Level Stewardship scheme (OELS) 137, 140
Organisation for Economic Co-operation and Development (OECD) 171–172
organochlorine (OC) 34
organophosphate (OP) 34–35
Orthoptera 21, 22

pesticides 32–33, 216–218
    carbamates 34
    development 33–35, 39
    dichlorodiphenyltrichloroethane (DDT) 23, 34
    fungicides 35
    herbicides 35
    integrated pest management (IPM) 3, 16, 39–41, 62–64, 144–145, 218–221
        adoption 53–55, 131–135
        agri-environmental policy 135–144
        biopesticides 75–76, 103–108, 219–221
        broad-acre crops 59–60
        crop protection 49–50, 53
        developing countries 60–62
        economic thresholds 48–49
        field vegetable crops 57–58
        greenhouse crops 55–57
        microbial agents 75–98
        natural compounds 51, 99–101, 103
        orchard crops 58
        population dynamics 41–49
    organochlorine (OC) 34
    organophosphate (OP) 34–35
    problems with 35
        broad-spectrum activity 36–37
        environmental persistence 36
        pest resurgence 37
        resistance 37–38
        secondary pests 37
        toxicity 38
    pyrethroids 34, 100
    regulation 5–8, 72–73, 108–116, 148–149
        biopesticides 172–174
        CRD 6, 8, 9, 135, 149–153, 154–158, 161, 173–174, 179–194, 208–209
        EU 5, 80, 158–169, 182
        innovation 8–10, 187–192
        OECD 171–172
        policy networks 177–195
        retail governance 198–199
        UK 6, 8, 149–158
        USA 8, 169–171
        WHO 171–172
    safety 32, 38
    tax 167–168
    withdrawal of 39, 160
Pesticides Safety Directorate (PSD) *see* Chemicals Regulation Directorate (CRD)
Phasmarhabditis hermaphrodita 92
pheromones 102–103, 107–108, 120–121
Phytophthora 28
    infestans 30–31, 48
    palmivora 97–98
plant breeding 51–52
plant pathogens 25–26
    bacteria 29
    fungi 26–28
    microbial control 92–93
        antagonism 93–95
        induced host resistance 95–97

plant pathogens *continued*
   nematodes 30
   oomycetes 28–29
   Phytophthora infestans 30–31, 48
   vectors 21–23, 24
   viruses 29, 86–88
      vectors 24
Pochonia chlamydosporia 95
policy networks 177–178, 194–195
   biopesticides 178–179
   change 182–187
   components 179–182
   institutional reform 193–194
   regulatory innovation 187–191
      pressures 191–192
political science 10–11
   complex systems 12–13
   experimentation 11–12
   individualistic fallacy 13–15
   Oedipus effect 11
population dynamics 41–42
   competition 43–44
   economic thresholds 48–49
   environmental conditions 43, 45
   life history strategies 42–43
   predation 44
   size increase
      alien species 45–48
      agricultural disturbance 44–45
      emerging pests 48
      environmental conditions 45
      resource concentration 44
portfolio economies 132–133
potato
   Colorado potato beetle 23, 105–106
   Phytophthora infestans 30–31, 48
predation 44
probenazole 96–97
production 76, 107
Pseudomonas 94
Puccinia
   chondrillina 98
   graminis f.sp. *tritici* 48
push–pull strategy 108
pyrethroids 34, 100
pyrrolizidine alkaloids (PAs) 107

regulation 5–8, 72–73, 108–116, 148–149
   biopesticides 172–174

CRD 6, 8, 9, 135, 149–151, 161, 173–174
   biopesticides scheme 157–158
   cost 158
   efficacy testing 156–157
   enforcement 154–156
   institutional reform 193–194
   maximum residue limits 152–153
   minor use crops 151–152
   off-label approvals 152
   policy networks 179–182
   regulatory innovation 187–192
   retailers 208–209
EU 5, 80, 158–159
   biopesticides 166
   climatic zones 161–162, 164–166
   Denmark 167–168
   directive 91/414 159–160, 162–163
   EPPO 161
   EFSA 159–160
   loss of active ingredients 160
   Netherlands 168–169
   policy networks 182
   reform 162–169
   shortcomings 160–162
   Sustainable Uses Directive 163–164
   Water Framework Directive 167
innovation 8–10, 187–191
   pressures 191–192
OECD 171–172
policy networks 177–178, 194–195
   biopesticides 178–179
   change 182–187
   components 179–182
   institutional reform 193–194
   regulatory innovation 187–192
retail governance 198–199
UK 6, 8, 149–158
   see also Chemicals Regulation Directorate (CRD)
USA 8, 169–171
WHO 171–172
residues 205
resistance 37–38
retailers 181, 198–199
   CRD 208–209

customers 210–213
Food Safety Act (1990) 201
power 199–201
regulation 201–204, 210–214
    cost 209–210
    pesticide classification 205–207
    residues 205
    standards organizations 204–205
    supply chain 207–208
rice 96–97
risk 132
Rodolia cardinalis 79
Rural Economy and Land Use (RELU) programme 2–3

safety
    botanicals 120–121
    human 32, 38, 115–118
    microbial agents 116–120
    semiochemicals 120–121
sawflies 21
semiochemicals 101–103
    safety 120–121
Septoria tritici 39
single farm payment 136–137
skeleton weed 98
slugs 25
snails 25
soybean 87, 221
spider mites 104
Spinosad 100
spittlebugs 89
spruce sawfly 87–88
standards organizations 204–205
Steinernema 105
    carpocapsae 105
    feltiae 91–92
    riobrave 105
straight-chain lepidopteran pheromones (SCLPs) 103, 120–121
stranglervine 98
sugarcane 89
sustainable farming 1–3
Sustainable Uses Directive 163–164
systemic acquired resistance 95–97

thrips 21, 22
Thysanoptera 21

tomato 56, 58
Trichoderma 93
    harzianum 93

UK
    Advisory Committee on Pesticides (ACP) 4
    agri-environmental policy (AEP) 135–144
    Chemicals Regulation Directorate (CRD) 6, 8, 9, 135, 149–151, 161, 173–174
    biopesticides scheme 157–158
    cost 158
    efficacy testing 156–157
    enforcement 154–156
    institutional reform 193–194
    maximum residue limits 152–153
    minor use crops 151–152
    off-label approvals 152
    policy networks 179–182
    regulatory innovation 187–192
    retailers 208–209
    entomopathogenic nematodes 75
    Environmental Stewardship scheme 137–144
    Food Safety Act (1990) 201
    integrated pest management (IPM) adoption 140–144
    policy networks 177–178
    retailers 199–201
USA
    cotton aphid 89
    cottony cushion scale 79
    gypsy moth 90
    hemlock woolly adelgid 47
    integrated pest management (IPM) 135–136
    Neozygites species 81
    regulation 8, 169–171
    western corn rootworm (WCR) 46–47

Varroa destructor 24–25, 38
vectors 21–23, 24
virus 29, 86–88
    vectors 24

Water Framework Directive 167
weeds 31–32
    microbial control agents 97–98
western corn rootworm (WCR) 46–47
western flower thrips 22, 38, 104
wheat 39
whitefly 22
World Health Organization (WHO) 171–172

yield 1–2
    integrated pest management 53–55
    loss 19–20

Zygomycota 88–89

www.ingramcontent.com/pod-product-compliance
Lightning Source LLC
Chambersburg PA
CBHW061259200426
43515CB00037B/429